METHODS AND PROTOCOLS IN FOOD SCIENCE

Series Editor
Anderson S. Sant'Ana
University of Campinas
Campinas, Brazil

For further volumes:
http://www.springer.com/series/16556

Methods and Protocols in Food Science series is devoted to the publication of research protocols and methodologies in all fields of food science.

Volumes and chapters will be organized by field and presented in such way that the readers will be able to reproduce the experiments in a step-by-step style. Each protocol will be characterized by a brief introductory section, followed by a short aims section, in which the precise purpose of the protocol will be clarified.

Biopolymers

Encapsulation Methods

Edited by

Ana Silvia Prata

Departamento de Engenharia e Tecnologia de Alimentos, Faculdade de Engenharia de Alimentos, Universidade Estadual de Campinas (UNICAMP), Campinas, Brazil

Louise Emy Kurozawa

Departamento de Engenharia e Tecnologia de Alimentos, Faculdade de Engenharia de Alimentos, Universidade Estadual de Campinas (UNICAMP), Campinas, Brazil

Humana Press

Editors
Ana Silvia Prata
Departamento de Engenharia e Tecnologia de Alimentos
Faculdade de Engenharia de Alimentos
Universidade Estadual de Campinas (UNICAMP)
Campinas, Brazil

Louise Emy Kurozawa
Departamento de Engenharia e Tecnologia de Alimentos
Faculdade de Engenharia de Alimentos
Universidade Estadual de Campinas (UNICAMP)
Campinas, Brazil

ISSN 2662-950X ISSN 2662-9518 (electronic)
Methods and Protocols in Food Science
ISBN 978-1-0716-5229-9 ISBN 978-1-0716-5230-5 (eBook)
https://doi.org/10.1007/978-1-0716-5230-5

This Humana imprint is published by the registered company Springer Science+Business Media, LLC, part of Springer Nature.
The registered company address is: 1 New York Plaza, New York, NY 10004, U.S.A.

Preface to the Series

Methods and Protocols in Food Science series is devoted to the publication of research protocols and methodologies in all fields of food science. The series is unique as it includes protocols developed, validated and used by food and related scientists as well as theoretical basis are provided for each protocol. Aspects related to improvements in the protocols, adaptations and further developments in the protocols may also be approached.

Methods and Protocols in Food Science series aims to bring the most recent developments in research protocols in the field as well as very well established methods. As such the series targets undergraduate, graduate and researchers in the field of food science and correlated areas. The protocols documented in the series will be highly useful for scientific inquiries in the field of food sciences, presented in such way that the readers will be able to reproduce the experiments in a step-by-step style.

Each protocol will be characterized by a brief introductory section, followed by a short aims section, in which the precise purpose of the protocol is clarified. Then, an in-depth list of materials and reagents required for employing the protocol is presented, followed by a comprehensive and step-by-step procedures on how to perform that experiment. The next section brings the do's and don'ts when carrying out the protocol, followed by the main pitfalls faced and how to troubleshoot them. Finally, template results will be presented and their meaning/conclusions addressed.

The *Methods and Protocols in Food Science* series will fill an important gap, addressing a common complain of food scientists, regarding the difficulties in repeating experiments detailed in scientific papers. With this, the series has a potential to become a reference material in food science laboratories of research centers and Universities throughout the world.

Campinas, Brazil ***Anderson S. Sant'Ana***

Preface to the Series

Preface

Encapsulation techniques are used in different fields, and there are numerous reviews in the literature addressing them, either individually or as general reviews. While the food sector has restrictions on raw materials compared to the pharmaceutical sector, on the other hand, raw materials originating from the agricultural sector (biopolymers, essential oils) can be used to innovate products in other areas such as agriculture, environmental applications, and materials, with more sustainable products. The opportunity to compile techniques more appropriate for the food sector aims to make available, in one place, the details of experimental practices for those who are beginning work with microencapsulation.

In this Protocols book, the selected Encapsulation Methods are described in general terms, with a focus on a combination of raw materials to enable an initial laboratory production of classic particles. The first step in particle development is to understand, for the desired application, what role encapsulation will play. Will it help mask flavor or alter some other sensory property? Or is the goal to provide protection during product processing, storage, or application/ingestion? Will it serve to modify some physical property that is important for the intended application, such as flowability, mixing, or prevention of interactions? These questions help to understand which characteristics the particles need to have—such as size, solubility, and distribution of the active compound—which will impact the release profile.

Such characterizations were organized into a single chapter (**Chapter 16**). In this chapter, a comparative table between encapsulation methods and the general properties of the particles obtained was also produced. With the definition of these characteristics and the required behavior during application, one can choose the encapsulation method which, starting from the same raw material, can result in particles with completely different properties. For example, starches can result in water-soluble and dry particles if produced by spray drying (**Chapter 4**), fluidized bed (**Chapter 1**), or if modified to produce porous starches (**Chapter 2**); in hydrogels, if they are gelled (**Chapters 8 and 9**); or in dry and insoluble particles if produced by thermoplastic extrusion without expansion (**Chapter 3**). Furthermore, it is possible to combine techniques, such as Layer-by-Layer (**Chapter 10**) applied to hydrogel particles, to associate the advantages of two techniques and produce particles with specific properties.

With this book, it is intended that classical techniques serve as a foundation so that brilliant scientists can develop new products, propose new applications, or create new encapsulation techniques.

Campinas, Brazil ***Ana Silvia Prata***
Campinas, Brazil ***Louise Emy Kurozawa***

Contents

Preface to the Series . . . . *v*
Preface . . . . *vii*
Contributors . . . . *xi*

1 Hot Melt Coating in Fluid Bed System 1
Bruna Barbon Paulo and Kaciane Andreola

2 Porous Starch for Probiotic Encapsulation 13
Ana Silvia Prata

3 Production of Nonexpanded Starch Particles Containing Essential Oils for Agro-Food Applications: A Detailed Protocol 19
Frank L. Romero-Orejon and Ana Silvia Prata

4 Microencapsulation of Hydrophilic and Hydrophobic Compounds by Spray Drying 25
Pedro Renann Lopes França and Louise Emy Kurozawa

5 Microencapsulation of Bioactive Compounds by Spray-Chilling Technique 37
Fernanda Ramalho Procopio, Marluci Palazzolli da Silva-Padilha and Miriam Dupas Hubinger

6 Solvent Evaporation 51
Gabriel Pedroso De Lima Alexandre and Ana Silvia Prata

7 Nanoencapsulation of Lipid Particles 59
Thaís J. Silva, Fernanda L. Lüdtke and Ana Paula B. Ribeiro

8 Internal and External Ionic Gelation 75
Maria Clara Guimarães and Ana Silvia Prata

9 Microencapsulation of Lipophilic Compounds by Complex Coacervation 81
Jorge A.D. Duarte, Nathalia D. Gonçalves and Ana Silvia Prata

10 Microencapsulation of Oils and Lipophilic Bioactive Compounds in Multilayer O/W Emulsions 89
Pedro Renann Lopes de França and Louise Emy Kurozawa

11 Pickering Emulsion and Encapsulation 99
Francisco E.R. Paes and Carolina S.F. Picone

12 Nanoprecipitation 107
Ana Silvia Prata

13 Pressurized CO_2 Antisolvent Coprecipitation of Bioactive Compounds in Polymeric Materials 113
Julia T. Paula, Ana P. Da Fonseca Machado and Ana C. Aguiar

14 Bioactive Compounds Vehiculated by Oleogels 121
Matheus A.S. Roman and Rosiane L. Cunha

15 In Situ Polymerization 133
Gisele Ongmayeb and Ana Silvia Prata

16 Interfacial Polycondensation. 137
Gisele Ongmayeb and Ana Silvia Prata

17 General Protocol for Particles Characterization. 143
Ana Silvia Prata

Subject Index. 161

Contributors

ANA C. AGUIAR • *Department of Food Science and Technology, Luiz de Queiroz College of Agriculture, University of São Paulo (ESALQ/USP), Piracicaba, SP, Brazil*

GABRIEL PEDROSO DE LIMA ALEXANDRE • *Departamento de Engenharia e Tecnologia de Alimentos (DETA), Faculdade de Engenharia de Alimentos, Universidade Estadual de Campinas (UNICAMP), Campinas, Brazil*

KACIANE ANDREOLA • *Institute of Food Technology, Campinas, Brazil*

ROSIANE L. CUNHA • *Departamento de Engenharia e Tecnologia de Alimentos, Faculdade de Engenharia de Alimentos, Universidade Estadual de Campinas (UNICAMP), Campinas, Brazil*

JORGE A. D. DUARTE • *School of Agricultural, Livestock and Environmental Sciences - ECAPMA, National Open and Distance University (UNAD), CEAD Ocana, Ocana, Colombia*

ANA P. DA FONSECA MACHADO • *Department of Rural and Animal Technology (DTRA), State University of Southwest Bahia (UESB), Itapetinga, Bahia, Brazil*

PEDRO RENANN LOPES DE FRANÇA • *Departamento de Engenharia e Tecnologia de Alimentos, Faculdade de Engenharia de Alimentos, Universidade Estadual de Campinas (UNICAMP), Campinas, Brazil*

NATHALIA D. GONÇALVES • *Departamento de Engenharia e Tecnologia de Alimentos, Faculdade de Engenharia de Alimentos, Universidade Estadual de Campinas (UNICAMP), Campinas, Brazil*

MARIA CLARA GUIMARÃES • *Departamento de Engenharia de Agronegócios, Escola de Engenharia Industrial Metalúrgica de Volta Redonda, Universidade Federal Fluminense (UFF), Volta Redonda, RJ, Brazil*

MIRIAM DUPAS HUBINGER • *Departamento de Engenharia e Tecnologia de Alimentos, Faculdade de Engenharia de Alimentos, Universidade Estadual de Campinas (UNICAMP), Rua Monteiro Lobato, Campinas, Brazil*

LOUISE EMY KUROZAWA • *Departamento de Engenharia e Tecnologia de Alimentos, Faculdade de Engenharia de Alimentos, Universidade Estadual de Campinas (UNICAMP), Campinas, Brazil*

FERNANDA L. LÜDTKE • *CEB—Centre of Biological Engineering, University of Minho, Braga, Portugal; LABBELS—Associate Laboratory, Braga/Guimarães, Portugal*

GISELE ONGMAYEB • *GinnOv Consulting, Nantes, France*

FRANCISCO E. R. PAES • *Departamento de Ciência de Alimentos e Nutrição, Faculdade de Engenharia de Alimentos, Universidade de Campinas (UNICAMP), Campinas, SP, Brazil*

JULIA T. PAULA • *Postgraduate Program in Materials Science, Institute of Exact and Earth Sciences, Federal University of Mato Grosso, Barra do Garças, MT, Brazil*

BRUNA BARBON PAULO • *Symrise Flavor and Fragrances Ltd., SP, Brazil*

CAROLINA S. F. PICONE • *Departamento de Engenharia e Tecnologia de Alimentos, Faculdade de Engenharia de Alimentos, Universidade de Campinas (UNICAMP), Campinas, Brazil*

ANA SILVIA PRATA • *Departamento de Engenharia e Tecnologia de Alimentos, Faculdade de Engenharia de Alimentos, Universidade Estadual de Campinas (UNICAMP), Campinas, Brazil*

FERNANDA RAMALHO PROCOPIO • *Embrapa Instrumentação (EMBRAPA), Rua XV de Novembro, São Carlos, SP, Brazil*

ANA PAULA B. RIBEIRO • *Departamento de Engenharia e Tecnologia de Alimentos, Faculdade de Engenharia de Alimentos, Universidade Estadual de Campinas (UNICAMP), Campinas, Brazil*

MATHEUS A. S. ROMAN • *Centro de Ciências da Natureza, Universidade Federal de São Carlos, Buri, Brazil*

FRANK L. ROMERO-OREJON • *Departamento de Engenharia e Tecnologia de Alimentos, Faculdade de Engenharia de Alimentos, Universidade Estadual de Campinas (UNICAMP), Campinas, Brazil*

THAÍS J. SILVA • *Center for Natural Sciences, Federal University of São Carlos (UFSCar), Lagoa do Sino Campus, Buri, Brazil*

MARLUCI PALAZZOLLI DA SILVA-PADILHA • *Departamento de Engenharia e Tecnologia de Alimentos, Faculdade de Engenharia de Alimentos, Universidade Estadual de Campinas (UNICAMP), Campinas, Brazil*

Check for updates

Chapter 1

Hot Melt Coating in Fluid Bed System

Bruna Barbon Paulo and Kaciane Andreola

Abstract

Hot-melt coating is a solventless technique that enables coat particles and protects the active material against external degradations, modifies the active release and/or sensory perception mainly for food and pharmaceutical products. This process can be performed using fluidized beds, where the particles are suspended in a bed through a fluid, and melted materials are atomized onto those particles and solidified subsequently. Complete coverage of the surface is achieved during several passes of particles through the coating zone. The knowledge of the thermal and rheological behavior of coating material, as well as the physicochemical properties of the particles, is essential for a successful hot-melt coating process. The coated particle can also be characterized to validate the coating performance. A reliable and comparable protocol for the coating process in fluidized bed is proposed in this chapter.

Keywords Solvent-free technology, Melt-coatings, Lipid-based excipients, Substrates

1 Introduction

Coating is a widely known process in the chemical, pharmaceutical, and food industries, commonly employed to achieve specific product performance characteristics [1, 2]. Conventionally, most coating formulations are solvent-based, which must be evaporated during and after coating. Nonetheless, solvent-free coatings, such as hot-melt coating, have been increasingly explored, as they can mitigate issues related to solvent use, including exposure risks, supply issues, and presence of residual solvent in the product [1, 3, 4].

Hot-melt coating process is performed with a substance in a molten state [4, 5] that solidifies after contact with the core material under temperatures higher than its solidification point. This kind of coating is mainly applied to modify the release of a certain substrate, drug, or active compound [6–13], to provide protection against physical or chemical degradation [14, 15], or to mask undesirable tastes [16, 17].

Ana Silvia Prata and Louise Emy Kurozawa (eds.), *Biopolymers*, Methods and Protocols in Food Science, https://doi.org/10.1007/978-1-0716-5230-5_1,

Hot-melt coating can be performed using two main types of equipment: pan coaters for larger particles [18, 19] or fluidized beds for smaller ones [5, 6, 11, 12, 20, 21]. Fluidized beds are vertical chambers in which an airstream is forced to move upward with enough velocity to suspend the particles. Coating agents are delivered through the spray nozzle located at the top or bottom of the column.

In general, the hot-melt coating process in fluidized bed consists of four steps. The first one involves setting up equipment and material preparation. In the second step, the substrate is heated, and the coating agent is melted and maintained at a temperature that ensures it remains in its molten state. The third one consists of dispersing the coating agent onto the substrate's surface. Cooling occurs during the final stage, enabling the lipid droplets to solidify into a uniform film coating [1].

This chapter describes a protocol with some strategies for operating a fluidized bed, for achieving pre-selection of coating materials, and for specific characterizations.

2 Materials

1. Core material: Cellulose microcrystalline (MCC, size 300 μm): 500 g (*see* **Note 1**).
2. Coating material: Carnauba wax (CW, melting point 80°C): 500 g (*see* **Note 2**).
3. DSC apparatus, aluminum pans.
4. Rheometer equipped with an oscillatory shear mode and a temperature control system (e.g., Peltier plate, oven, or liquid cooling).
5. Goniometer apparatus (*see* **Note 3**).
6. Compressed plate of MCC (*see* **Note 4**).
7. A set of standard sieves of various required microns.
8. Vibrational system.
9. Balance +/− 0.01 g.

2.1 Description of Equipment

Figure 1 illustrates the general schematic of the hot-melt coating process in fluid bed systems, showing the different steps involved in the top-spray method.

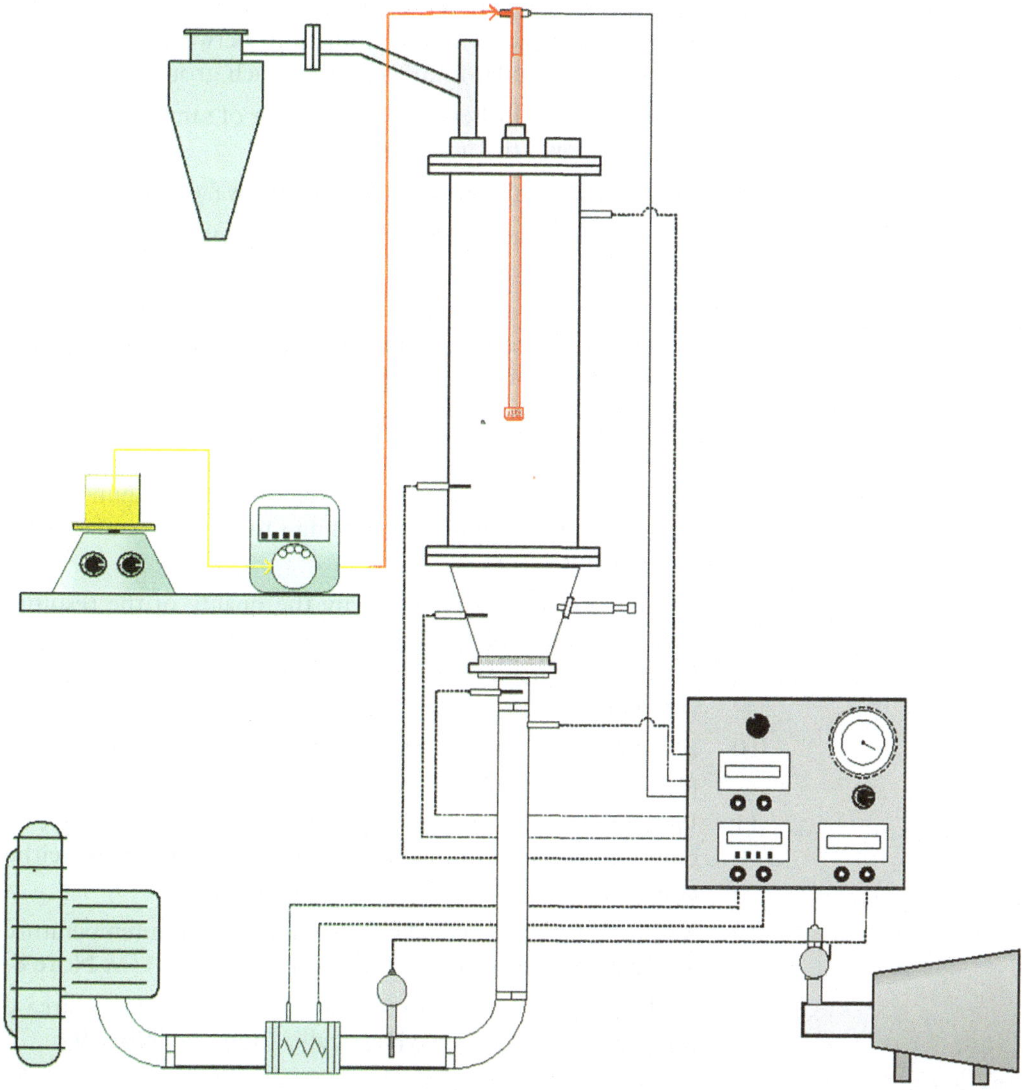

Fig. 1 Schematic of hot-melt coating process in a top-spray fluid bed. (1) compressor; (2) ball valve; (3) anemometer; (4) electrical resistance, (5) centrifugal blower, (6) control panel; (7) differential pressure transmitter; (8) thermocouples; (9) air distributor plate; (10) sampling tube; (11) fluid bed chamber; (12) bi-fluid spray nozzle; (13) peristaltic pump; (14) molten coating material; (15) heating plate; (16) cyclone(Source: the author)

3 Methods

3.1 Pre-Selection of the Coating Materials

3.1.1 Melting and Crystallization Temperatures

1. Accurately weigh the pans, lids, and the sample (*see* **Note 5**).

2. Weigh both parts together pans + lids and write down the total weight. Try to find combinations between pans and lids to result in a similar total weight for each group.
3. Accurately weight between 6 and 7 mg of sample in the pans and note down the final mass.
4. Close the lid and compact it on the press (*see* **Note 6**).
5. Consider the sealed pan and check the total weight (sample + pan + lid). Note down.
6. An empty reference set of pan+lid must be employed; the weight should correspond to the pan+lid of the sample (variation in the fourth decimal case).
7. Switch on the calorimeter and wait at least 30 min to equilibrate.
8. Adjust the pressure of the oxygen (50 mL min^{-1}) and nitrogen gas supplies according to the manufacturer's recommendations.
9. Set the run parameters to verify the quality of the resulting peak: heating-cooling cycles from 0 to 150°C at 2°C min^{-1}, with isothermal processes of 15 min at final temperatures (*see* **Note 7**).
10. Fill the empty pan weight and the sample weight into the measuring program.
11. Place the sample pan in the right part of the cell, and place the empty reference pan in the left part. Make sure that neither cup is touching the wall of the measuring cell.
12. Melting (Tm) and crystallization (Tc) transition temperatures are defined at the peak top temperatures (Tpeak) (*see* **Note 8**).
13. Enthalpy of the transitions is calculated through the area of the peak (ΔHfusion or ΔHcrystallization) (*see* **Note 9**).

3.1.2 Evaluation of Rheologic Behavior

1. Choose the Cone-plate geometry (60 mm, 1.59°) (*see* **Note 10**).
2. Load a known amount of the melted wax into the geometry, with a little excess, ensuring uniformity, degassing to remove bubbles, and trimming excess material.
3. Switch on the rheometer and wait at least 1 h to equilibrate.
4. Set the gap size (e.g., 0.5–2 mm for soft materials).
5. Calibrate the temperature control system and ensure the instrument is thermally equilibrated.
6. Set the run parameters: shear rate ranging from 0 to 300 s^{-1} at 85 and 100°C (*see* **Note 11**).
7. Lower the measuring cone to the trim gap width. Ensure that the gap is filled with sample; If air bubble formation or any gap

is observed, discard the measurement result, and repeat the experiment.

8. Fit the data from the third flow curve to a generic model (Herschel–Bulkley): $\tau = \tau_0 + k(\gamma)^n$, where τ_0 is the initial shear stress (Pa), k is the consistency index (Pa.s^n), γ is the shear rate (s^{-1}), and n is the behavior index (−);
9. Apparent viscosity (η) can be determined at 100 s^{-1} as a result of the adjust to the power law equation.

3.1.3 Oscillatory Shear Measurements

1. Perform a strain sweep test at a fixed temperature to determine the linear viscoelastic region (LVR). Use strain values well within the LVR for subsequent measurements.
2. Perform a frequency sweep test at the same temperature to select a representative frequency (e.g., 1 Hz or 10 rad/s).

Oscillatory Temperature Ramp Test

1. Program the temperature range of interest (e.g., 60°C to 150°C) (*see* **Note 11**).
2. Heating rate: 10°C min^{-1}.
3. Cooling rate: 10°C min^{-1}.
4. Apply a constant strain within the LVR and a constant angular frequency (e.g., 1 Hz).
5. Set the measurement interval for collecting data (e.g., every 0.1–1°C).
6. Record the elastic modulus (G′), loss modulus (G″), and phase angle ($\delta = \tan^{-1}(G''/G')$) as a function of temperature.
7. Monitor for transitions (e.g., glass transition, gelation, melting, or cross-linking) indicated by changes in G′ and G″.
8. Plot G′ and G″ versus temperature.
9. Identify critical temperatures: Crossover point: Where G′ = G″, indicative of a sol-gel or gel-sol transition, and Inflection points: Reflect phase transitions like glass transition (Tg) or melting (Tm).
10. Repeat the test with replicate samples to ensure reproducibility.
11. If needed, perform multiple heating and cooling cycles to assess thermal stability.

3.1.4 Adhesion

1. Switch on the goniometer and wait at least 30 min to equilibrate (*see* **Note 12**).
2. Adjust the parameters of the camera.
3. Define the conditions of temperature and relative humidity.

4. Generate a drop of the lipidic coating material by a 1 mL syringe using a hook needle, both previously cleaned, at temperature enough to enhance the melting.
5. Deposit the droplet on a flat surface of the substrate or a water droplet.
6. Capture the images using the camera at defined number of pictures for time.
7. Visually measure the contact angle between the coating material and the substrate and analysis the results using the appropriate software.
8. At least twenty measurements are needed for reproducible results.

3.2 Hot-Melt Coating Process

3.2.1 Equipment Set Up: Fluid Bed

1. Assemble the fluidized bed apparatus with provisions for airflow control, bed pressure measurement (using a differential manometer or a differential pressure transmitter), atomization nozzle, and air distributor plate (*see* **Note 13**).
2. Insert a defined mass of particles in the fluid bed (*see* **Note 14**).
3. Gradually increase airflow through the bed from zero to a point well beyond the expected minimum fluidizing velocity and terminal velocity. Both velocities can be estimated by:

$$umf = \frac{\mu \ Remf}{\rho g \ dp}$$

$$Remf = \left[(28.7)^2 + 0.0494 \ \frac{(dp^3 \ \rho g(\rho p - \rho g)g)}{\mu^2}\right]^{0.5} - 28.7$$

$$ut^* = \left[\frac{18}{dp^*} + \frac{2.335 - 1.744 Øs}{(dp^*)^{0.5}}\right]^{-1}$$

$$dp^* = dp\left[\frac{\rho g(\rho p - \rho g)g}{(\mu)^2}\right]^{1/3}$$

where ρp and ρg are the density of particle and air (kg/m³), respectively; dp is the particle diameter (m); g is the acceleration of gravity (9.8 m/s²), μ is the viscosity of air (Pa.s), Øs is the sphericity of particles.

4. Record the air velocity and corresponding pressure drop at each step.
5. Plot ΔP as a function of air velocity and identify:
   - A plateau region, where the initial point of this region corresponds to umf.
   - The region where particles are carried out of the bed corresponds to the terminal velocity (ut).
   - From these conditions, establish the fluidizing air velocity to be employed in the coating process (*see* **Note 15**).

3.2.2 Coating Process

1. Pour the pre-determined amount of the coating material into the shape stainless steel plate pre-heated to melt the coating material, enabling the appropriate pumping.
2. Set the fluidizing air velocity as previously determined.
3. Switch on the compressor and set the atomizing air as determined previously based on preliminary tests.
4. Determine the position for the nozzle: top-spray, bottom-spray, or tangential spray (*see* **Note 16**).
5. Transfer the coating material by a pump through a jacketed tube to a jacketed fluid atomizer nozzle.
6. It is recommended to insulate the pumping coating material, which is melted (*see* **Note 17**).
7. Switch the electrical resistance control to the heat position. The temperature set must be set below the melting point of the coating material ($T<T_m$) to avoid its melting inside the equipment.
8. Insert an analytical balance to record the mass of the coating material delivered to the particles (core material) inside the equipment. Begin recording when the solution fills the tube up to the nozzle.
9. Switch on the pump and spray the slurry onto the particles at the appropriate pressure and flow rate. Observe the movement of the particles (fluid dynamic behavior). If the movement begins to change (indicating the initiation of defluidization), stop the atomization and wait for the movement to return to its original state.
10. When the coating process starts, measure the total time in minutes (t) until the spraying is finished.
11. Stop the pumping after the level of coating to be achieved (see methodology below).
12. Fluidize the coated substrate for 2–3 min, allowing good, uniform application.

13. Once this is completed, allow the blower to cool for 3–5 min before switching off the system.
14. Remove the samples from the fluid bed and package as appropriate.
15. Clean the equipment using soap detergent. Ensure all the residues are cleaned out.
16. Rinse with clean water and dry with a cloth.
17. Switch the mains to the off position on the machine.

3.3 Characterization of the Coated Substrate

The particle should be evaluated for size and density, as described in Chap. 17. Specific characterization for the hot-melt coating process is described here.

3.3.1 Coating Efficiency (CE) and Process Yield (PY)

1. Weigh the amount of particles (core materials) placed in the fluid bed ($M_{\text{uncoated particles}}$).
2. After ending the coating process, record the amount of coating material sprayed onto particles (M_{coating}), the time taken to coat the particles, and the amount of coated particles recovered particles ($M_{\text{coated particles}}$).
3. The theoretical mass of coating material added to the equipment can also be defined from the process parameters and the physicochemical properties of the coating material: $M_{coating} = Q_{at} \times \rho_{coating} \times t$.
4. Coating efficiency (CE) is given by:

$$CE = \frac{M_{coated} - M_{uncoated}}{M_{coating}} \times 100$$

5. Process yield (PY) is given by:

$$PY = \frac{M_{coated}}{\left(M_{uncoated} + M_{coating}\right)} \times 100$$

3.3.2 Agglomeration Index (Agl)

1. Select a standard set of Tyler sieves, at least five different mesh sizes/openings.
2. Place the smallest required sieve size at the bottom of the collecting pan, followed by the biggest sieve size until the required number of sieve sizes has been reached.
3. Weigh the total amount of coated particles (m total) in a glass beaker.

4. Transfer the sample into the top sieve and close with the sieve lid.
5. Swiftly shake and tap the sieves for 10 min at least.
6. Remove the lid and slowly decant remaining sample from the sieve to the glass beaker.
7. Weigh the particles reminiscent in each sieve (m sieve).
8. Determine the % distribution of powder for each sieve as a ratio of $\frac{m\ sieve}{m\ total} \times 100$.
9. The sieves with a higher amount indicate the range of the mean diameter of the particles.
10. Agglomeration index (Agl) is defined as:

$$Agl = \frac{M_{agl}}{M_{total}} \times 100$$

where M_{agl} is the amount of particles with diameters 3× higher than the uncoated substrate; Weigh the total of coated substrate (Mfinal substrate).

4 Notes

1. Hot-melt coating can be applied to fine powders [11, 5], granules [12, 21–23], pellets [3, 6], and spherules [4].
2. Commonly used excipients for hot-melt coating are lipid substances such as waxes [3, 20, 24], vegetable oils [3, 24], and their derivatives such as polyoxylglycerides [3, 6], fatty acids [5, 9], and partial glycerides [4, 6, 12, 22].
3. Tensiometer can equally be used for measuring contact angle.
4. A plate of MCC can be done using a hydraulic press. Firstly, weigh a known amount of the core material and deposit it on a metal-bracket composed by two plates with delimited edges. Insert this metal bracket in the hydraulic press, under pre-set values of temperature and pressure after preliminary tests [25].
5. Do not touch the sample or crucibles by hand!
6. For this analysis, do not perforate the lid. Be sure that the lid is well pressed.
7. These parameters were set specifically for carnauba wax. For other materials, other parameters can be set accordingly to their range of melting and solidification point.

8. Melting is an endothermic event, and crystallization is an exothermic one. In the case that the baseline is not horizontal, changes in the run parameters must be done.
9. Preferably, hot-melt coating materials should demonstrate physicochemical stability at temperatures up to 200°C, a narrow melting range, and no softening prior to melting for a specified crystallization, and a melting point between 75°C and 80°C the minimum drying temperature is 40–60°C [3, 11, 20, 24].
10. This geometry is indicated for fluids from very low to high viscosities. Generally, the viscosity of the molten lipid excipient is less than 300 mPa s at 80°C [1].
11. This temperature is chosen based on the melting point of the coating material. For avoiding issues of phase transition, two temperatures above the melting point can be selected to validate the results.
12. A tensiometer can equally be used.
13. Adequate air distribution is ensured by using a perforated plate, selected according to the particle size. A configuration with orifices of 0.001 mm in diameter, arranged in a triangular pattern with a spacing of 0.0085 mm, can be considered [26, 27].
14. Core material should present a bulk density between 0.5 and 1.0 g/mL, a particle size distribution as narrow as possible, and physicochemical stability at drying temperatures [24].
15. The fluid dynamic behavior of the particles may change as the coating is deposited. Therefore, a fluidizing air velocity higher than the minimum fluidization velocity should be used to avoid bed collapse, with the lower limit being at least 1.1 times the minimum fluidization velocity, considering the increase in particle size and mass throughout the process [27–30].
16. Fluidized beds configured for top, bottom (spouted bed and Wurster coater), or tangential spraying are considered for the coating process [5, 6, 11, 12, 20, 21]. Bottom spray is commonly used, and if necessary, a Wurster tube can be used. For top spray (see Fig. 1), adjust the distance between the nozzle and the particles. The nozzle should be positioned as close as possible to the particles to minimize the distance between the lipid droplets and the substrate.
17. Tubes carrying the molten lipid from the reservoir to the spray nozzle can be heated to prevent crystallization of the lipid by using oil or vapor.

References

1. Bose S, Bogner RH (2007) Solventless pharmaceutical coating processes: a review. Pharm Dev Technol 12:115–131
2. Werner SR, Jones JR, Paterson AH, Archer RH, Pearce DL (2007) Air-suspension particle coating in the food industry: part I—State of the art. Powder Technol 171(1):25–33
3. Achanta AS, Adusumilli PS, James KW, Rhodes CT (1997) Development of hot melt coating methods. Drug Dev Ind Pharm 23(5):441–449
4. Barthelemy P, Laforet JP, Farah N, Joachim J (1999) Compritol® 888 ATO: an innovative hot-melt coating agent for prolonged-release drug formulations. Eur J Pharm Biopharm 47(1):87–90
5. Kulah G, Kaya O (2011) Investigation and scale-up of hot-melt coating of pharmaceuticals in fluidized beds. Powder Technol 208(1):175–184
6. Sinchaipanid N, Junyaprasert V, Mitrevej A (2004) Application of hot-melt coating for controlled release of propranolol hydrochloride pellets. Powder Technol 141(3):203–209
7. Padsalgi A, Bidkar S, Jadhav V, Sheladity D (2008) Sustained release tablet of theophylline by hot-melt wax coating technology. Asian J Pharm 2(3):26–29
8. Sakarkar DM, Jaiswal SB, Dorle AK, Deshmukh VN (2009) Application of cow ghee as hot-melt coating agent in the design of sustained-release pellets. Int J PharmTech Res 1:1167–1172
9. Patil AT, Khobragade DS, Chafle SA, Ujjainkar AP, Umathe SN, Lakhotia CL (2012) Development and evaluation of a hot-melt coating technique for enteric coating. Braz J Pharm Sci 48:69–77
10. Jeong KH, Woo HS, Kim CJ, Lee KH, Jeon JY, Lee SY, Kang JH, Lee S, Choi YW (2015) Formulation of a modified-release pregabalin tablet using hot-melt coating with glyceryl behenate. Int J Pharm 495(1):1–8
11. Salar-Behzadi S, Corzo C, Lopes DG, Meindl C, Lochmann D, Reyer S (2020) Novel approach for overcoming the stability challenges of lipid-based excipients. Part 2: application of polyglycerol esters of fatty acids as hot melt coating excipients. Eur J Pharm Biopharm 148:107–117
12. Milanovic A, Aleksic I, Ibric S, Parojcic J, Cvijic S (2018) Hot-melt coating with Precirol ATO 5 in a fluidized-bed apparatus: application of experimental design in the optimization of process parameters. J Drug Deliv Sci Technol 46:274–284
13. Huynh VTT, Lacerda SP, Espitalier F, Beyssac E, Ré MI (2024) Effect of talc and vitamin E TPGS on manufacturability, stability and release properties of trilaurin-based formulations for hot-melt coating. Int J Pharm 653:123866
14. Achanta AS, Adusumilli PS, James KW, Rhodes CT (2001a) Hot melt coating water sorption behavior of excipient films. Drug Dev Ind Pharm 27:241–250
15. Chen H, Shi S, Liu A, Tang X (2010) Combined application of extrusion–spheronization and hot-melt coating technologies for improving moisture-proofing of herbal extracts. J Pharm Sci 99:2444–2454
16. Mittal B, Kidney D, Sy E, Chu J (2003) Taste masking of aspirin using hot-melt coating approach. AAPS Pharm Sci Tech 3(S1):000720
17. Patil A, Chafle S, Khobragade D, Umathe S, Avari J (2011) Evaluation of hot melt coating as taste masking tool. Int Res J Pharm 2:169–172
18. Sakarkar DM, Dorle AK, Mahajan NM, Sudke SG (2013) Design of sustained release pellets of ferrous fumarate using cow ghee as hot-melt coating agent. Int J Pharm Investig 3(3):151–156
19. Sudke SG, Sakarakar DM (2013) Design and characterization of enteric coated pellets of aspirin using hot-melt coating technique. Int J Pharma Res Rev 2(3):1–10
20. Müller MG, Lindner JA, Briesen H, Sommer K, Foerst P (2018) On the properties and application of beeswax, carnauba wax and palm fat mixtures for hot melt coating in fluidized beds. Adv Powder Technol 29:781–788
21. Lopes DG, Garsuch V, Becker K, Paudel A, Stehr M, Zimmer A, Salar-Behzadi S (2016) Improving the granule strength of roller compacted ibuprofen sodium for hot-melt coating processing. Int J Pharm 510(1):285–295
22. Faham A, Prinderre P, Farah N, Eichler KD, Kalantzis G, Joachim J (2000) Influence of Compritol® 888 Ato and granule size on theophylline release. Drug Dev Ind Pharm 26:167–176
23. Knezevic Z, Gosak D, Hraste M, Rausl D, Khan MZI (2009) Application of hot-melt coating process for designing a lipid based controlled release drug delivery system for highly aqueous soluble drugs. Chem Pharm Bull 57(5):464–471
24. Jannin V, Cuppok Y (2013) Hot-melt coating with lipid excipients. Int J Pharm 457:480–487
25. Andreola K (2017) Agglomeration of concentrated rice protein in a fluidized bed: in-line monitoring of particle size. Ph.D. thesis. Department of Process Engineering, School of Chemical Engineering, University of Campinas, Campinas, São Paulo, Brazil
26. Silva CAM, Taranto OP (2015) Real-time monitoring of gas–solid fluidized-bed granulation

and coating process: evolution of particle size, fluidization regime transitions, and psychometric parameters. Dry Technol 33:1929–1948
27. Andreola K, Silva CAM, Taranto OP (2018) Agglomeration process of rice protein concentrate using glucomannan as binder: in-line monitoring of particle size. Chem Eng Res Des 135:37–51
28. Machado VG, Hirata TAM, Menegalli FC (2014) Agglomeration of soy protein isolate in a pulsed fluidized bed: experimental study and process optimization. Powder Technol 254:248–255
29. Paulo BB, Andreola K, Taranto OP, Ferreira AD, Prata AS (2019) Coating approach for a phase change material (PCM). Powder Technol 341:147–156
30. Custodio GR, de Souza LFG, Nitz M, Andreola K (2020) A protein powder agglomeration process using açaí pulp as the binder: an analysis of the process parameters. Adv Powder Technol 31:3551–3561

Chapter 2

Porous Starch for Probiotic Encapsulation

Ana Silvia Prata

Abstract

Physical, chemical, or enzymatic modification in conventional starches results in granules with several pores in the micrometric range, transforming the conventional starch in a porous structure serving as a suitable support for entrap active substances. Owing to its surface area, biodegradability, non-toxicity, and hydrophilic and hydrophobic cavities, porous starch (PS) has been extensively used as the delivery agent of oils, vitamins, enzymes, and microorganisms in food applications. This protocol describes the encapsulation of a probiotic in PS prepared through enzymatic hydrolysis, the mostly used modification.

Keywords Starch, Enzymatic hydrolysis, Probiotic strain, Adsorption

1 Introduction

Starch is an abundant, inexpensive, degradable, non-polluting, and renewable resource. Porous starch is a product that retains the granular structure of native starch and forms a porous honeycomb surface whose void volume can reach up to 50% of the volume of starch granules, which has a large specific surface area and a rich microporous structure. Because crystalline and amorphous characteristics are present, hydrophilic–hydrophobic properties are developed in their cavities, and several target core materials can be physically adsorbed in the pores and cavities without any covalent bonding [1].

In the food industry, porous starch is often used as a carrier for coloring agents, spices, flavor agents, sweeteners, and some unstable substances like unsaturated fatty acid glycerides, vitamins, β-carotene, and lycopene [2].

The porous structure can be obtained via physical, chemical, enzymatic, and combined synergistic methods, and the porosity depends on processing conditions and starch source. Understanding and controlling these parameters allows customization for specific applications, from pharmaceutical drug delivery systems to enhancing texture and moisture retention in food products.

Ana Silvia Prata and Louise Emy Kurozawa (eds.), *Biopolymers*, Methods and Protocols in Food Science,
https://doi.org/10.1007/978-1-0716-5230-5_2,

In this chapter, the focus will be given in obtaining porous starch by enzymatic method, and for this reason, maize starch is preferable because its structure facilitates a higher enzymatic hydrolysis susceptibility [3]. Several enzymes and combinations of them have been employed to deform starch, such as α-amylase, amyloglucosidase, glycogen branching enzyme, cyclodextrin-glycosyltransferase [4–7]. The protocol is described for entrapment of a probiotic bacteria and the modification conditions described for pancreastin and native maize starch.

2 Materials

- Microorganisms: *Lactobacillus plantarum* 299v (from commercial source) (*see* **Note 1**).
- de Man, Rogosa, and Sharpe (MRS) broth.
- MRS agar.
- Phosphate-buffered saline (PBS).
- 0.2 M sodium phosphate buffer (pH 6.3).
- Peptone water (0.1% w/v).
- Native maize starch (22.2% amylose).
- Pancreatin (*see* **Note 2**).
- 5 M HCl.
- 1 M NaOH.
- Sachet of anaerobic gas generator (e.g., Anaero Gen).
- Oxgall bile salt (3% w/v).
- Ethanol (for washing starch).
- LIVE/DEAD BacLight bacterial viability kit.
- Maltose colorimetric kit determination.
- Sterile tubes and falcon tubes.
- Petri plates (100 × 15 mm).

2.1 Equipment

- Incubator.
- Anaerobic jar or system.
- Centrifuge (capable of 4000 × *g* and 10,000 × *g*).
- Vacuum oven (40°C).
- Freeze dryer.
- Freezer.
- Water bath (37°C).
- Magnetic stirrer and stir bars.
- Orbital shaker (capable of 250 × *g* and 600 × *g*).

- Dry block heater (60°C).
- Autoclave (for media sterilization).
- pH meter.
- Analytical balance (accurate to mg scale).
- Confocal laser scanning microscope.

3 Methods

Sterilize all instruments, solutions, and media prior to using them for plating procedures.

3.1 Preparation of Agar Plates

1. Autoclave the agar at 121°C for 15 min.
2. After autoclaving, cool the agar to between 45°C and 50°C prior to pouring the plates.
3. Add 15 mL of agar per plate (100 × 15 mm), aiming to reach 0.3 cm of thickness.
4. Store the plates at room temperature for about 24 h to dry the agar.
5. Once dried, plates may be used or refrigerated in closed bags or containers until required.

3.2 Probiotic Culture

1. Make a serial dilution taking 1 mL of the commercial source of *L. plantarum* 299v with 9 mL of sterile peptone water (0.1% w/v) or phosphate-buffered saline.
2. Repeat the procedure six times for each last dilution created (dilutions from 10^{-1} to 10^{-6}).
3. Take 1 mL of the dilutions and spread it in MRS agar, spreading the most dilute samples first (*see* **Notes 3** and **4**).
4. Avoid disturbing plates for 10 to 20 min after spreading.
5. Incubate at 37°C for 72 h under static conditions in an incubator.
6. Randomly select the colonies streaked on agar plates.
7. Inoculate the strain in MRS broth.
8. Incubate at 37°C for 24 h under static conditions in an incubator.
9. Harvest cells by centrifugation at 4400 × *g* for 10 min.
10. Wash and centrifuge the pellet twice in sterile 0.2 M sodium phosphate buffer (pH 6.3).
11. Resuspend in the same buffer to obtain a concentration of approximately 10^{10} CFU/mL.

3.3 Preparation of Porous Starch

1. Prepare a 5% (w/v) maize starch suspension in phosphate-buffered saline (PBS).

2. In 50 mL Falcon tubes, add the enzyme at 0.5 U per mg starch. Prepare controls without enzyme.
3. Incubate these tubes in a water bath maintained at 37°C, continuously stirring with a magnetic stirrer bar at 250 × *g* for 30 min.
4. Recover the starch by centrifuging the tubes at 4000 × *g* for 5 min.
5. Wash residue three times with ethanol to remove soluble sugars and residual enzyme.
6. Transfer the starch material to Petri plates and vacuum-oven dry at 40°C, overnight.
7. Store dry porous starch for further encapsulation in plastic bags or hermetic pots.
8. Determine the degree of hydrolysis (DH) of starch through the quantification of the released maltose through a colorimetric maltose assay kit (*see* **Note 5**).

3.4 Encapsulation of L. plantarum

1. Mix 6 mL bacterial culture with 2.0 g porous starch in a sterile tube.
2. Stir on an orbital shaker at 600 × *g* for 3 h.
3. Let it stand for 2 h to settle.
4. Remove supernatant carefully with a pipette.
5. Transfer the sediment to a Petri plate and pre-cool in a freezer (−20°C) for 4 h.
6. Freeze-dry overnight.

For the coating material:

1. Prepare gelatinized starch by heating 2% (w/v) native starch in water for 15 min, until it forms a gel.
2. Cool to room temperature.
3. Mix freeze-dried powder with gelatinized starch at 3 mL gel/g starch.
4. Recrystallize at −20°C overnight, then freeze-dry for 24 h.
5. Store microcapsules at 4°C in sterile tubes.

3.5 Viable Bacteria Count

1. Suspend 0.10 g microcapsules in 0.9 mL peptone water (0.1% w/v) + pancreatin (0.5 U/mg starch).
2. Shake at 600 × *g*, 15 min.
3. Serial dilute in peptone water (0.1% w/v) from 10^{-1} to 10^{-6}.
4. Spread 0.1 mL onto MRS agar.
5. Incubate at 37°C, 36 h under anaerobic conditions produced with a sachet of anaerobic gas generator.

6. Count CFU on plates with 15–300 colonies.

   The viability loss can be calculated by the equation:

- Reduction = log N_0 − log N
- where N_0 = initial count, N = count after stress.

3.6 Acid and Bile Salt Resistance

3.6.1 Acid Test

1. Use MRS broth pH 2.0 (adjusted with 5 M HCl, autoclaved).
2. Add 0.90 mL broth to 0.10 g microcapsules or 0.10 mL free cells.
3. Incubate at 37°C for 1 h.
4. Neutralize with 1 M NaOH.
5. Determine viable cells as described above.

3.6.2 Bile Test

1. Prepare MRS + 3% Oxgall bile salt (pH 5.8).
2. Add 0.90 mL to 0.10 g capsules or 0.10 mL free cells.
3. Incubate at 37°C for 2 h and 4 h.
4. Determine viable cells as described above.

3.7 Heat Resistance

1. Mix 0.10 g capsules or 0.10 mL free cells with 0.90 mL phosphate buffer (pH 6.3).
2. Incubate at 60°C for 15 min (dry block heater).
3. Cool to 23 ± 2°C in a water bath.
4. Determine viable cells as described above.

3.8 Viable cells

1. Stain ~10 μg powder with 10 μL LIVE/DEAD BacLight stain.
2. Incubate 30 min in the dark.
3. Observe at 488 nm excitation, 550 nm emission, 63× oil objective in a Confocal Laser Scanning miccroscope (*see* **Note 6**).

   The particle should be evaluated for size and density, as described in Chap. 17.

4 Notes

1. If necessary, confirm strain identity by 16S rDNA sequencing [8].
2. Other enzymes can be used, such as pancreatic α-amylase and fungal α-amylase

3. Since some bacteria rapidly attach to the agar surface, the inoculum should be spread soon after it is applied.
4. A reusable glass or metal spreader should be flame sterilized by dipping in alcohol (such as 70% isopropyl or ethanol), shaking off the excess alcohol, and igniting the residue.
5. DH is estimated to be 18%–22%, but it can reach up to 41%, if the incubation time is extended to 120 min. For some strains, there was no significant difference in terms of protection and loading capacity [9]. Other determination techniques can be found [10].
6. This assay serves to identify the location of the viable cells into the particle.

References

1. Wang X, Yuan Y, Yue T (2015) The application of starch-based ingredients in flavor encapsulation. Starch Stärke 67(3–4):225–236. https://doi.org/10.1002/star.201400163
2. Qian J, Yang H, Mo C, Chen Y, Zhao C (2022) Preparation of porous starch from native starch by using fungal amylase and evaluation of its adsorption property on natural pharmacodynamic compounds. Polym Adv Technol 33(7):2244–2253. https://doi.org/10.1002/pat.5675
3. Dhital S, Shrestha AK, Gidley MJ (2010) Relationship between granule size and in vitro digestibility of maize and potato starches. Carbohydr Polym 82(2):480–488. https://doi.org/10.1016/j.carbpol.2010.05.018
4. Davoudi Z, Azizi MH, Barzegar M (2022) Porous corn starch obtained from combined cold plasma and enzymatic hydrolysis: microstructure and physicochemical properties. Int J Biol Macromol 223(Part A):790–797. https://doi.org/10.1016/j.ijbiomac.2022.11.058
5. Dura A, Rosell CM (2016) Physico-chemical properties of corn starch modified with cyclodextrin glycosyltransferase. Int J Biol Macromol 87:466–472. https://doi.org/10.1016/j.ijbiomac.2016.03.012
6. Purwitasari L, Wulanjati MP, Pranoto Y, Witasari LD (2023) Characterization of porous starch from edible canna (Canna edulis Kerr.) produced by enzymatic hydrolysis using thermostable α-amylase. Food Chem Adv 2:100152. https://doi.org/10.1016/j.focha.2022.100152
7. Zhu Q, Tang J, Yao S, Feng J, Mi B, Zhu W, Chen Q, Liu D, Xu E (2023) Controllable structure of porous starch facilitates bioactive encapsulation by mild gelatinization. Food Hydrocolloids 145:109135. https://doi.org/10.1016/j.foodhyd.2023.109135
8. Drancourt M, Bollet C, Carlioz A, Martelin R, Gayral JP, Raoult D (2000) 16S ribosomal DNA sequence analysis of a large collection of environmental and clinical unidentifiable bacterial isolates. J Clin Microbiol 38:3623–3630. https://doi.org/10.1128/jcm.38.10.3623-3630.2000
9. Li H, Ho VTT, Turner MS, Dhital S (2016) Encapsulation of lactobacillus plantarum in porous maize starch. LWT 74:542–549. https://doi.org/10.1016/j.lwt.2016.08.019
10. Bilskey SR, Olendorff SA, Chmielewska K, Tucker KR (2020) A comparative analysis of methods for quantitation of sugars during the corn-to-ethanol fermentation process. SLAS Technol 25(5):494–504. https://doi.org/10.1177/2472630320908253

Chapter 3

Production of Nonexpanded Starch Particles Containing Essential Oils for Agro-Food Applications: A Detailed Protocol

Frank L. Romero-Orejon and Ana Silvia Prata

Abstract

The encapsulation of volatile and thermolabile compounds, such as essential oils, is essential to preserve their bioactivity and broaden their use in food and agricultural applications. Hot-melt extrusion (HME) offers a scalable and efficient strategy for producing starch-based carriers capable of protecting and controlling the release of these compounds. Here, we present a detailed, step-by-step protocol for the encapsulation of oregano (*Origanum vulgare*), thyme (*Thymus vulgaris*), and savory (*Satureja montana*) essential oils using a co-rotating twin-screw extruder. The procedure covers feed material preparation, oil incorporation, extrusion under controlled conditions (temperature profile: 60–90–80–70°C; screw speed: 300 rpm; feed rate: 4 kg/h), and post-processing steps including drying and storage. Additionally, we outline methods for characterizing the final product in terms of moisture content, bulk density, expansion ratio, and encapsulation performance. This reproducible framework provides a practical foundation for developing starch-based functional carriers with potential applications in sustainable food preservation and crop protection.

Keywords Hot melt extrusion, Encapsulation, Essential oils, Starch-based matrices, Volatile compounds, Controlled release

1 Introduction

Extrusion is an efficient technology for producing solid molecular dispersions. This process converts raw materials into a uniform product by forcing them through a die under controlled conditions of temperature, feed rate, pressure, screw speed, and moisture [1, 2].

Encapsulation by extrusion is widely applied in the pharmaceutical, food, and agricultural industries to protect sensitive compounds and enable their controlled release [3–5]. Depending on the application, the active material may include vitamins, essential oils, flavors, or plant growth-promoting bacteria [6–8].

Ana Silvia Prata and Louise Emy Kurozawa (eds.), *Biopolymers*, Methods and Protocols in Food Science,
https://doi.org/10.1007/978-1-0716-5230-5_3,

In food and agricultural contexts, polysaccharides and protein-based polymers are frequently used as encapsulating matrices [1]. Plasticizers such as water or glycerol may be required to improve processing and enhance the physical and mechanical properties of the extrudate [9]. Suitable polymers should exhibit a glass transition temperature between 50–180°C, low hygroscopicity, thermal stability under extrusion conditions, good miscibility with the core material, and nontoxicity, since they are used in relatively high amounts [10, 11].

Successful encapsulation requires careful selection of both the wall material and the active compound, as well as optimization of operational conditions (screw configuration, feed rate, screw speed, residence time, energy input, barrel temperatures). These parameters define the final particle characteristics, including expansion and density.

The aim of this protocol is to provide a reproducible procedure for the encapsulation of essential oils by hot-melt extrusion, using starch-based matrices to produce nonexpanded particles suitable for food and agricultural applications.

2 Materials and Equipment

2.1 Materials

- Wall materials: Corn starch with low amylose content (1.8%, Amisol 4000) or medium amylose content (28%, Amisol 3408) (Ingredion, Brazil) (*see* **Note 1**).
- Core materials: Thermolabile compounds such as oregano essential oil (*Origanum vulgare*), thyme essential oil (*Thymus vulgaris*), or savory essential oil (*Satureja montana*) (*see* **Note 2**).
- Plasticizer: Distilled water.

2.2 Equipment

- Analytical balance.
- Infrared moisture balance.
- Planetary mixer.
- Stainless-steel sieve (1.00 mm mesh size).
- Twin-screw extruder (Fig. 1), featuring:
    - Length-to-diameter (L/D) ratio: 29:1.
    - Screw configuration: Low shear (*see* **Note 3**).
    - Die outlet diameter: 2.5 mm.
- Forced Air Laboratory Oven.

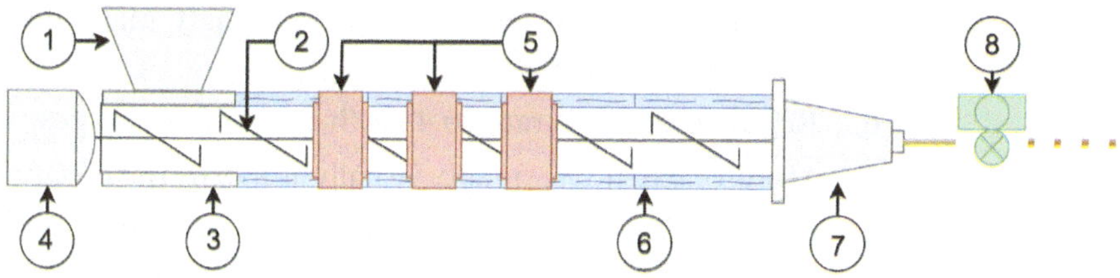

Fig. 1 Schematic of Twin-Screw Extruder. (1) Feeder; (2) Rotating screw: metal rod with a varying diameter and a helical fillet around it; (3) Barrel; (4) motor; (5) electrical resistance, (6) Cooling system (water bath or chilled air); (7) Die head; (8) Cutter

2.3 Equipment Description

Figure 1 illustrates the general schematic of the extrusion equipment. Temperature and pressure sensors are installed in the extruder barrel to monitor operational conditions. The process involves feeding, melting, mixing, and shaping of materials under controlled conditions to obtain the desired encapsulates.

3 Methods

3.1 Preparation of Feed Material

1. Determine the moisture content of the starch to be used using an infrared moisture balance.
2. Adjust starch moisture to 28% (w/w) by gradually adding distilled water while mixing in a planetary mixer at 80 rpm.
3. Seal the hydrated starch in polyethylene bags and store for 12 h at 8°C to ensure uniform hydration.
4. On the day of extrusion, add the essential oil to the hydrated starch to achieve a final concentration of 5% (w/w, wet basis) (*see* **Note 4**).
5. Homogenize manually until the essential oil is thoroughly incorporated, ensuring the mixture is visually uniform and lump-free.
6. Pass the mixture through a 1.00 mm sieve to break up agglomerates and prevent feeding problems.

3.2 Perform the Encapsulation Process

1. Ensure the extruder is clean, correctly assembled, and in operational condition before use (*see* **Note 5**).
2. Feed the prepared mixture into the extruder.
3. Switch on the extruder and preheat to the operational temperatures.
4. Set barrel temperatures across four zones (from feed to die): 60°C, 90°C, 80°C, and 70°C.

5. Adjust the screw speed at 300 rpm and maintain constant throughout the process.
6. Regulate the feed rate to 4 kg/h to avoid blockages.
7. Monitor extrudate quality visually as it exits the die(*see* **Note 6**).
8. Collect the extruded particles and separate them immediately to prevent sticking during cooling.

3.3 Extruder Shutdown and Postextrusion Handling

1. Gradually reduce screw speed to avoid material buildup in the barrel.
2. Once stopped, turn off heating elements and cooling system.
3. Allow the barrel to cool before performing surface cleaning.

3.4 Drying and Storage Conditions

1. Allow the extruded particles to cool to room temperature.
2. Dry the cooled particles in a forced-air oven at 70°C for 3 h to achieve a final moisture content below 15%.
3. Store the dried encapsulates in airtight containers in a clean, dry place at room temperature.

3.5 Characterization of Encapsulates

1. The particle should be evaluated for size, moisture content and bulk density, as described in Chap. 7.
2. Expansion Ratio: Measure extrudate diameter at 20 random points with a caliper. Calculate as the ratio of mean extrudate diameter to the die diameter.

4 Notes

1. Medium-amylose starch (28%) typically gelatinizes between 62°C and 80°C, which should be considered when defining temperature profiles.
2. Core material can be an active ingredient, drug, or bioactive compound. For essential oils, it is important to know their chemical composition. A high content of phenolic compounds, such as carvacrol (in oregano oil) or thymol (in thyme oil), is suggested as this may confer greater thermal stability during extrusion.
3. Low-shear screw configuration used: 3:60/30, 1:42/21, 1:28/14, 1:21/21, 1KB 90, 1:21/21, 1:28/14, 8:20/10, 1:28/14, 1KB 90, 2:28/14, 1:14/14, 12:20/10 (A = length in mm; B = pitch; KB = kneading block, with adjacent crest angle).
4. Formulations generally consist of 70%–90% coating material and 5%–20% core material.

Table 1
List of encapsulation of different bioactive compounds by extrusion

Bioactive compound	Wall material	Extrusion conditions	Reference
Quercetin	Carnauba wax, Shella, and Zein	Barrel temp.: 80 to 90°C Screw speed: 100 rpm	[12]
Cinnamaldehyde, eugenol, and nonanoic acid	Corn starch/b-cyclodextrin	Barrel temp.: 60, 120, 110, 100°C Screw speed: 100 rpm Die: 7 mm	[13]
Thyme essential oil	Starch/water	Barrel temp.: 60, 90, 100, 90°C Screw speed: 300 rpm Die: 3 mm	[14]
Azospirillum brasilense strain Ab-V5,	Cassava starch, Glycerol, crystal sugar, powdered skim milk, yeast extract, PBS	Barrel temp.: 120°C Screw speed: 200 rpm Die: 4 mm	[15]

5. Verify that motors, control panels, and the cooling system are functioning. Cold water is recommended for higher heat transfer efficiency.
6. Check extrudate quality: diameter (~1.0 mm), transparency, hardness, and stickiness. If unsuitable, adjust extrusion conditions accordingly.

Previous studies (Table 1) show that extrusion formulations are usually prepared before processing, and that careful control of temperature and pressure is required to minimize degradation of bioactive compounds.

References

1. Bamidele OP, Emmambux MN (2021) Encapsulation of bioactive compounds by "extrusion" technologies: a review. Crit Rev Food Sci Nutr 61:3100–3118
2. Velloso CCV, Lopes MM, Badino AC et al (2023) Exploring the roles of starch for microbial encapsulation through a systematic mapping review. Carbohydr Polym 306:120574
3. Reque PM, Brandelli A (2021) Encapsulation of probiotics and nutraceuticals: applications in functional food industry. Trends Food Sci Technol 114:1–10
4. Saberi Riseh R, Hassanisaadi M, Vatankhah M et al (2023) Encapsulating biocontrol bacteria with starch as a safe and edible biopolymer to alleviate plant diseases: a review. Carbohydr Polym 302:120384
5. Smaoui S, Hlima H Ben, Ben Braïek O et al (2021) Recent advancements in encapsulation of bioactive compounds as a promising

technique for meat preservation. Meat Sci 181:108585
6. Ali M, Cybulska J, Frąc M et al (2023) Application of polysaccharides for the encapsulation of beneficial microorganisms for agricultural purposes: a review. Int J Biol Macromol 244:125366
7. Li X, Hong X, Shuai S et al (2024) A review of hot melt extrusion technology: advantages, applications, key factors and future prospects. J Drug Deliv Sci Technol 98:105884
8. Xu Y, Yan X, Zheng H et al (2024) The application of encapsulation technology in the food industry: classifications, recent advances, and perspectives. Food Chem X 21:101240
9. Guo Y, Qiao D, Zhao S et al (2021) Starch-based materials encapsulating food ingredients: recent advances in fabrication methods and applications. Carbohydr Polym 270:118358
10. Jani R, Patel D (2015) Hot melt extrusion: an industrially feasible approach for casting orodispersible film. Asian J Pharm Sci 10:292–305
11. Lang B, McGinity JW, Williams RO (2014) Hot-melt extrusion – basic principles and pharmaceutical applications. Drug Dev Ind Pharm 40:1133–1155
12. Khor CM, Ng WK, Kanaujia P et al (2017) Hot-melt extrusion microencapsulation of quercetin for taste-masking. J Microencapsul 34:29–37
13. Kollengode ANR, Hanna MA (1997) Cyclodextrin complexed flavors retention in extruded starches. J Food Sci 62:1057–1060
14. Maia JD, Corte R La, Martinez J et al (2019) Improved activity of thyme essential oil (Thymus vulgaris) against Aedes aegypti larvae using a biodegradable controlled release system. Ind Crops Prod 136:110–120
15. Marcelino PRF, Milani KML, Mali S et al (2016) Formulations of polymeric biodegradable low-cost foam by melt extrusion to deliver plant growth-promoting bacteria in agricultural systems. Appl Microbiol Biotechnol 100:7323–7338

Check for updates

Chapter 4

Microencapsulation of Hydrophilic and Hydrophobic Compounds by Spray Drying

Pedro Renann Lopes De França and Louise Emy Kurozawa

Abstract

Spray drying is the most employed microencapsulation technique, which can produce microparticle powders containing various compounds of interest through the atomization and drying of a feed solution. This process is fast, economical, highly scalable, and easy to perform. The resulting powders generally provide effective protection for different compounds and applicability in various food matrices. However, proper execution of the process and careful handling of both wall and core materials are necessary to ensure successful microencapsulation. Thus, this chapter presents a practical protocol for microencapsulation of hydrophilic and hydrophobic compounds by spray drying, including suggestions of analysis to characterize the feed solutions.

Keywords Wall materials, Core material, Protection, Encapsulation

1 Introduction

Spray drying is typically considered a dehydration process, but it can also be applied to encapsulate a desired material in a protective polymeric matrix [1]. It is a continuous process that consists of a one-step transformation of a liquid feed solution into dried particles by a combined process of atomization to form small droplets, and rapid drying [2].

Several microencapsulation techniques have been developed over the years, but spray drying remains the most commonly used technology in the food industry due to its low cost and readily available equipment [3]. It is one of the oldest encapsulation methods, having been used since the 1930s to produce the first encapsulated flavors [4]. Despite its benefits, the process includes some disadvantages, such as the exposure to a heated air stream, which can lead to the degradation of thermolabile compounds. Additionally, the process generally requires that wall materials possess good water solubility,

Ana Silvia Prata and Louise Emy Kurozawa (eds.), *Biopolymers*, Methods and Protocols in Food Science, https://doi.org/10.1007/978-1-0716-5230-5_4,

low viscosity, and good flow properties, limiting the applicability of different polymers [5].

The aim of this chapter is to present a protocol for encapsulation of hydrophilic and hydrophobic compounds by spray drying, emphasizing the preparation and handling of different feed solutions using carbohydrates and proteins as wall materials. Additionally, this chapter will assist in selecting key process parameters for drying.

2 Materials

- Hydrophilic core material: some pigments [6], vitamins [7], aromas [8], and phenolic compounds [9]
- Hydrophobic core material: oils [10], vitamins [11], pigments [12], and essential oils [13]
- Wall material: carbohydrates (gum arabic, maltodextrin, modified starches, pectin, etc.) and/or proteins (whey protein, casein, some vegetable proteins, etc.) [14] (*see* **Note 1**).
- Fuming hydrochloric acid (37%).
- Sodium hydroxide (NaOH solution, 1 M) and Chloride acid (HCl solution, 1 M).
- Ultrapure water (conductivity = 18 MΩ·cm at 25°C).
- Analytical balance.
- Equipment for powder storage (desiccator or freezer).
- Magnetic stirrer with a magnetic bar.
- pH meter.
- Rotor-stator homogenizer.
- Spray dryer.
- Ultrasonic bath.

3 Description of Equipments

Figure 1 presents the general scheme of the spray drying process, showing the different steps involved in the process.

Spray dryer consists of different components that play crucial roles in the drying process [15]:

I. **Pump:** Typically, laboratory-scale equipment uses a peristaltic pump. The pump is responsible for transporting the feed solution to the atomization/drying chamber. It controls the flow rate of the liquid to ensure a constant feed during the process.

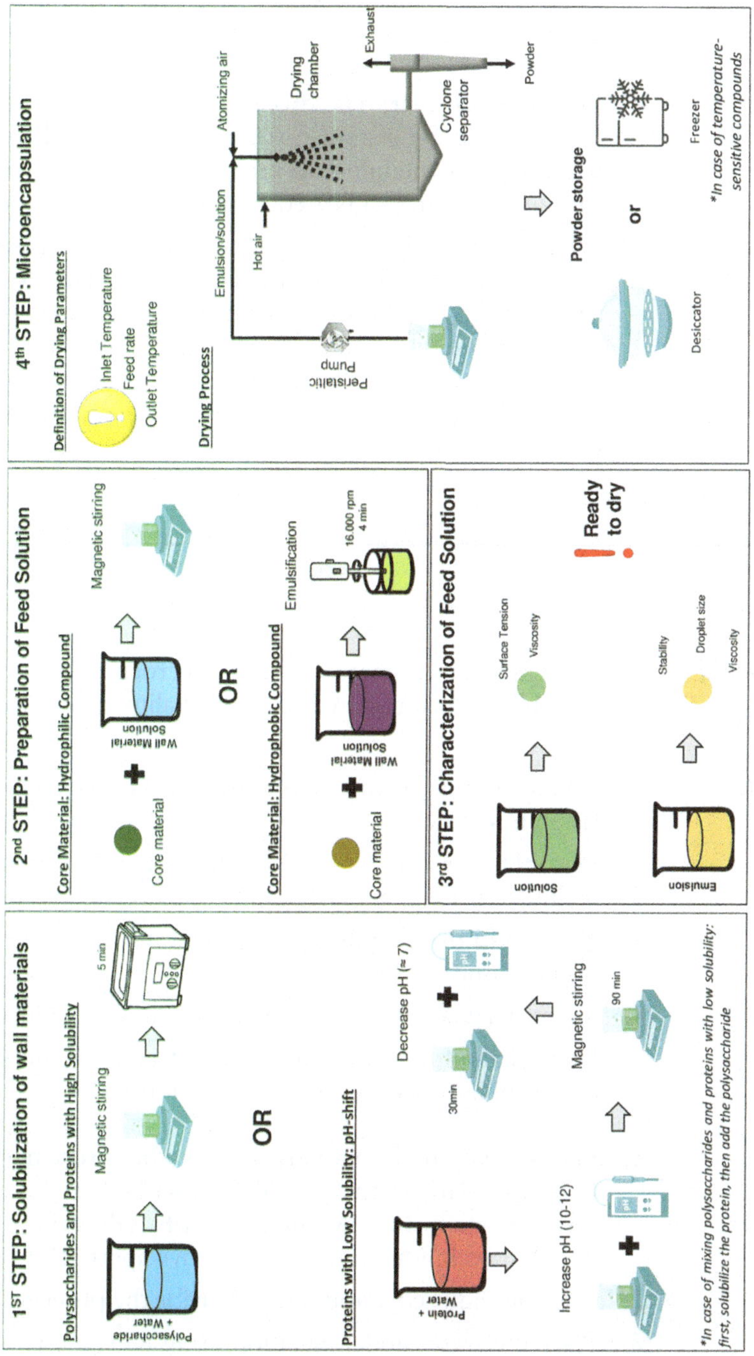

Fig. 1 General scheme for the microencapsulation by spray drying. (Source: the author)

II. **Atomizer Nozzle:** This part is critical for the entire process, as it converts the feed solution into small droplets. Generally, this process involves high-pressure atomization, where compressed air forces the liquid through the nozzle to form the droplets. Other systems include a rotary atomizer, which uses a rotating disk to create the droplets.

III. **Drying Chamber:** The droplets are released by the atomizer inside the drying chamber, where they are rapidly dried, forming powders containing the microparticles. This process occurs through the entry of heated air at the desired temperature.

IV. **Cyclone:** After drying, the airflow carries the formed powders from the drying chamber to the cyclone. Due to its unique shape, the cyclone uses centrifugal force to separate the powders from the air stream. The powders follow a downward flow, where they can be collected in a suitable container, while the heated air follows an upward flow and is released through an exhaust system.

4 Methods

4.1 Solubilization of Wall Materials

Regardless of the wall material used in microencapsulation, its complete solubilization is required for good quality of the formed powder (Fig. 1).

4.1.1 Carbohydrates

1. Add the required amount of water for the feed solution formulation into a beaker.
2. Place the water under agitation using a magnetic stirrer and a magnetic bar (*see* **Note 2**).
3. Slowly add the carbohydrate to the water, avoiding the formation of lumps.
4. Maintain agitation until complete dissolution (*see* **Notes 3 and 4**).
5. Place in an ultrasonic bath for 5 min to remove bubbles, which are mainly formed in very viscous solutions.

4.1.2 Proteins

1. Proteins with high solubility, such as whey protein, can be solubilized similarly to the previously described method. However, for proteins with low solubility, pH adjustment using the pH-shift technique can be an effective approach [16].
2. Add the required amount of water for the process to a beaker.
3. Place the water under agitation using a magnetic stirrer and a magnetic bar.
4. Slowly add the protein to the water, avoiding the formation of agglomerates, and stir until all the material is hydrated.

5. Monitor the pH, and with the aid of a Pasteur pipette, add NaOH solution to the mixture until it reaches pH 12.
6. Allow protein solubilization for 90 min.
7. Gradually reduce the pH to 7.0 by dripping HCl 1 M solution (*see* **Notes 5, 6, and 7**).
8. Maintain agitation for an additional 30 min.

4.2 Preparation of the Feed Solution: Hydrophobic Core Material

1. Immediately after preparing the solution of wall materials, add the core material (*see* **Notes 8, 9, and 10**).
2. Perform emulsification of the system for complete dispersion of the core material using a rotor-stator homogenizer at 16,000 rpm for 4 min (*see* **Note 11**).

4.3 Preparation of the Feed Solution: Hydrophilic Core Material

1. Immediately after preparing the solution of wall materials, disperse the core material into it (*see* **Notes 8 and 9**).
2. Using a magnetic stirrer and a magnetic bar, maintain stirring for approximately 30 min (*see* **Note 12**).

4.4 Characterization of the Feeding Solution

The following characteristics of feed solutions are fundamental to be understood, as they significantly impact the quality of the resulting powders. Additionally, for emulsions, analysis of oil droplet sizes and emulsion stability is included.

4.4.1 Viscosity

An optimal viscosity is required for atomization and drying. High viscosity can hinder pumping to the nozzle and the atomization process, resulting in larger droplet formation in the chamber dryer, which causes difficulties in water evaporation. Conversely, low viscosity may lead to low emulsion stability and structural breakdown during atomization, increasing the surface oil content [10]. In addition, low-viscosity solutions may produce very fine powders that are easily carried away by the drying air. Therefore, an optimal viscosity is required [17].

Recommended method: In this case, it is advisable to assess the rheological behavior of the emulsion using a rheometer, as this will provide information regarding the dependency of viscosity on shear rate, which is significant since the shear rate during atomization is high [18].

4.4.2 Surface Tension

Feeding solutions with high surface tension may impede droplet formation, leading to large particles with a highly polydispersed size distribution. Additionally, there may be significant migration of core material to the surface of the particles, resulting in low encapsulation efficiency. The formation of a fragile surface film may also occur, leading to unstable particles [19] (*see* **Note 13**).

Recommended methods: Measurement should be performed using a tensiometer, assessing the tension between the droplet and the air [6].

4.4.3 Emulsion Stability

A minimum stability of the emulsion is required for the spray drying process. Emulsions with low stability tend to have their structure compromised during the atomization process, leading to an increased surface oil content and consequently reducing the efficiency of encapsulation and the recovery of the powder [10].

Recommended methods: Turbiscan emulsion stability analysis [10] and the creaming index [20].

4.4.4 Emulsion Oil Droplet Size

Generally, smaller droplet sizes and a narrow size distribution are desired. Conversely, larger droplets may break during the atomization of the feed solution, leading to higher surface oil content. Furthermore, emulsions with elevated droplet sizes and high polydispersity tend to be more unstable [10].

Recommended methods: This can be measured using instruments such as the static light scattering [10] (droplet diameter > 1 μm), dynamic light scattering [21] (droplet diameter < 1 μm), and/or microscopy analyses [19].

4.5 Definition of Drying Parameters

The selection of drying parameters primarily depends on the formulation of the feeding solution and the spray dryer to be used. Therefore, it is essential to consider the desired characteristics of the powder (moisture content, particle size, etc.) and the stability of the core material. The main parameters that will affect the quality of the produced powders include the inlet air temperature, the feeding solution flow rate, and the outlet air temperature [1]. Generally, it is recommended to conduct preliminary tests and consult the literature for the definition of these parameters. However, below is a summary of how each parameter may influence the formation of powders.

4.5.1 Inlet Temperature

The inlet temperature is the most critical parameter to be set. An appropriate choice will promote effective drying and preservation of the compounds. Generally, this temperature can be set between 140°C and 220°C. Lower temperatures may result in moister powders. The determination of this parameter is typically based on the core material's resistance to heating; the higher the temperature, the easier it will be to remove water and form powders [1].

4.5.2 Outlet Temperature

The outlet temperature is not controlled during the drying process. The value obtained will result from a complex energy balance, primarily depending on the material feeding rate and inlet temperature. It is important to note that very high outlet temperatures lead to greater losses of thermolabile compounds and particles with moister interiors, due to the rapid formation of a rigid and dry layer on the particle surfaces [22].

4.5.3 Feeding Rate

The feeding rate will control the operational time. With this parameter, it is possible to regulate the residence time of each droplet within the drying chamber. Therefore, the feeding rate must be sufficient for the droplets to achieve an adequate degree of drying upon contacting the chamber walls. High feeding rates lead to the production of larger and moister particles, whereas low feeding rates result in the formation of smaller and drier particles [23, 24] (*see* **Note 14**).

4.6 Spray Drying

The production of microparticles will occur in the spray dryer (Fig. 1, fourth step microencapsulation), with the final process forming powders containing these microparticles.

1. Verify the operating conditions and assembly provided by the manufacturer to ensure proper equipment operation.
2. After assembling the spray dryer, set the air flow (according to the manufacturer's recommendation), inlet air temperature, and pump feeding rate (chosen by preliminary test or based on the literature).
3. Activate the ventilation system.
4. Turn on the heating system.
5. When the inlet air temperature reaches the desired value, turn on the pumping system (*see* **Note 15**), passing distilled water;
6. Wait the necessary time for the equipment to reach a steady-state operating condition (*see* **Note 16**).
7. With a fresh feeding solution/emulsion, connect it to the pumping system.
8. Initiate the drying operation and start powder production.
9. Monitor the atomization of feed solution into the dryer chamber during drying to ensure it is adequate, along with product drying and proper powder formation (*see* **Note 17**).
10. Once the entire solution has dried, stop the pumping and heating systems.
11. Keep the ventilation system until the outlet air temperature drops to ~50°C.
12. Turn off the ventilation system.
13. Collect the resulting powder in an amber glass container (*see* **Note 18**).
14. Storage conditions may vary; generally, protect against moisture by storing in a desiccator containing silica. If the material is sensitive to different environmental conditions, it is advisable to store it under freezing conditions in properly sealed containers for subsequent analysis (*see* **Note 19**).
15. Disassembly and clean the equipment and the nozzle carefully using appropriate cleaning accessories.

The particle should be evaluated for size, solubility (Reconstitution), and moisture content, as described in Chap. 17.

5 Notes

1. Wall Material Selection: The selection of the wall material should be made in advance, based on literature reviews [25]. Additives such as surfactants, emulsifiers, and antioxidants may also be considered to act in conjunction with wall materials to protect the active compounds [26, 27].
2. Solubilization of materials: If the material is not soluble at ambient temperature, it may be necessary to apply heat. However, maintaining the feeding solution at elevated temperatures may increase the degradation of thermolabile compounds.
3. pH control for carbohydrates: When using carbohydrates with limited solubility, such as chitosan, it is important to adjust the system's pH accordingly, as chitosan is only soluble in acidic conditions.
4. Dissolution time for carbohydrates: Some carbohydrates are easy to dissolve, such as maltodextrin, requiring approximately 1 h of stirring for complete solubilization. However, others, such as gums, pectin, or alginate, may require more stirring time, and it may be necessary to maintain the solution under agitation overnight.
5. pH adjustment: pH variation should be gradual to avoid protein precipitation. If the addition of HCl is causing a rapid decrease, reduce the concentration to 0.1 M.
6. pH and protein functionality: Generally, the pH is maintained at neutrality (around seven). However, care should be taken to keep this value away from the isoelectric point of the protein, as proteins exhibit low techno-functional properties near this point [10].
7. Sequential solubilization: In cases where proteins and carbohydrates are used together, if the protein has low solubility, it is recommended to first achieve its solubilization before adding the carbohydrate.
8. Core material concentration: It is important to note that high concentrations of core material can increase the compound load in the final product but may also reduce encapsulation efficiency [18].
9. Total solids content: typical working range between 20% and 60% [19]. Within this range, the viscosity, stability, and wall material-to-core material ratio generally exhibit favorable characteristics [18].
10. Interfacial activity agents: In this case, it is necessary for the feeding solution to contain an agent with interfacial activity,

such as proteins, surfactants, or Pickering-type emulsifiers, as these will help stabilize the emulsified system.

11. Homogenization methods: The method suggested in this protocol utilizes a rotor-stator homogenizer. However, these systems may not always produce emulsions with adequate droplet sizes. Additional homogenization methods, such as high-pressure homogenization, ultrasonic treatment, or microfluidization, may be employed to achieve kinetically more stable emulsions [28].
12. Co-encapsulation: in the case of co-encapsulating hydrophilic and hydrophobic compounds, the preparation should follow the method used for hydrophobic compounds, but with the hydrophilic compound dispersed in the aqueous phase.
13. Improving surface tension: In cases of high surface tension, surfactants such as proteins or surfactants may be added to improve this property and enhance powder quality.
14. In the case of laboratory-scale sprays, it is not recommended to work with large volumes of solution, as this can result in long drying periods, exposure of the active compounds to high temperatures, and low powder recovery. Volumes between 150 and 400 mL, depending on the set drying parameters, are sufficient to ensure a process within a reasonable time and powder recovery.
15. Pump system: The pump should be adjusted to the desired flow rate for the process.
16. Stabilization time in spray dryers: For laboratory-scale spray dryers, 30 min is usually sufficient for both inlet and outlet temperatures to stabilize. However, keep in mind that this is not a fixed parameter and will depend on the equipment used.
17. Monitoring material accumulation: It is essential to monitor the spray dryer process to determine whether there is an accumulation of undried material in the drying chamber, as this is an indication that atomization is not occurring adequately or that the chosen drying parameters are unsuitable. Therefore, check whether the nozzle is operating correctly; if not, consider revising the drying parameters.
18. Powder recovery: In some cases, there may be some powder partially stuck to the surface of the cyclone. If this occurs at the end of the process, the powder can be recovered with the help of a dry brush.
19. Storage conditions: If stored under freezing conditions, keep the powder samples in an ultra-freezer (−80°C) if possible, as at this temperature, chemical and biochemical reactions in the food matrix are halted. If an ultra-freezer is not available, a temperature of −18°C is sufficient [29].

References

1. Gharsallaoui A, Roudaut G, Chambin O et al (2007) Applications of spray-drying in microencapsulation of food ingredients: an overview. Food Res Int 40:1107–1121
2. Archana M, Kumar J, Chandra Bhainsa K (2021) Applicability of spray drying technique to prepare nano-micro carriers: a review. Nanoarchitectonics 3(1):33–45
3. Furuta T, Neoh TL (2021) Microencapsulation of food bioactive components by spray drying: a review. Drying Technol 39:1800–1831
4. Shahidi F, Han X (1993) Encapsulation of food ingredients. Crit Rev Food Sci Nutr 33:501–547
5. Yang M, Liang Z, Wang L et al (2020) Microencapsulation delivery system in food industry—challenge and the way forward. Adv Polym Technol 2020:1–14
6. Fernandes Almeida R, Gouveia Gomes MH, Kurozawa LE (2024) Enzymatic hydrolysis improves the encapsulation properties of rice bran protein by increasing retention of anthocyanins in microparticles of grape juice. Food Res Int 180:114090
7. Hu S, Ding Z, Zhang G et al (2022) Fabrication and spray-drying microencapsulation of vitamin C-loaded W1/O/W2 emulsions: influence of gel polymers in the internal water phase on encapsulation efficiency, reconstituted stability, and controlled release properties. LWT 170:114113
8. Sanchez-Reinoso Z, Osorio C, Herrera A (2017) Effect of microencapsulation by spray drying on cocoa aroma compounds and physicochemical characterisation of microencapsulates. Powder Technol 318:110–119
9. Da Silva Júnior ME, Araújo MVRL, Martins ACS et al (2023) Microencapsulation by spray-drying and freeze-drying of extract of phenolic compounds obtained from ciriguela peel. Sci Rep 13: 15222
10. França PRLD, Gomes MHG, Nascimento RF et al (2024) Microencapsulation of linseed oil in multilayer emulsions stabilized by carob protein hydrolysate and pectin. Food Hydrocolloids 150:109641
11. Bashir I, Wani SM, Bhat AA et al (2024) Effect of freeze drying and spray drying on physical properties, morphology and in vitro release kinetics of vitamin D3 nanoparticles. Powder Technol 432:119164
12. González-Peña MA, Lozada-Ramírez JD, Ortega-Regules AE (2021) Antioxidant activities of spray-dried carotenoids using maltodextrin-Arabic gum as wall materials. Bull Natl Res Cent 45:58
13. Gomes MHG, Kurozawa LE (2024) Rice protein hydrolysates as natural emulsifiers for an effective microencapsulation of orange essential oil by spray drying. Drying Technol 42:1791–1800
14. Keshani S, Daud WRW, Nourouzi MM et al (2015) Spray drying: an overview on wall deposition, process and modeling. J Food Eng 146:152–162
15. Selvamuthukumaran M (eds) (2019) Handbook on spray drying applications for food industries. CRC Press, Boca Raton
16. Sales ACA, De Oliveira RIB, Tavares GM (2024) Mixing oat proteins with lentil or faba bean proteins: how is the techno-functional behavior of the different protein systems affected by heat treatments?. Food Hydrocolloids 157:110461
17. Ré MI (1998) Microencapsulation by spray drying. Drying Technol 16:1195–1236
18. Galves C, Galli G, Miranda CG et al (2021) Improving the emulsifying property of potato protein by hydrolysis: an application as encapsulating agent with maltodextrin. Innov Food Sci Emerging Technol 70:102696
19. Mcclements DJ (2007) Critical review of techniques and methodologies for characterization of emulsion stability. Crit Rev Food Sci Nutr 47:611–649
20. Gomes MHG, Kurozawa LE (2020) Improvement of the functional and antioxidant properties of rice protein by enzymatic hydrolysis for the microencapsulation of linseed oil. J Food Eng 267:109761
21. Hong IK, Kim SI, Lee SB (2018) Effects of HLB value on oil-in-water emulsions: droplet size, rheological behavior, zeta-potential, and creaming index. J Ind Eng Chem 67:123–131
22. George S, Thomas A, Kumar MVP et al (2023) Impact of processing parameters on the quality attributes of spray-dried powders: a review. Eur Food Res Technol 249:241–257
23. Altay Ö, Köprüalan Ö, İlter I et al (2024) Spray drying encapsulation of essential oils; process efficiency, formulation strategies, and applications. Crit Rev Food Sci Nutr 64:1139–1157
24. Ziaee A, Albadarin AB, Padrela L et al (2019) Spray drying of pharmaceuticals and biopharmaceuticals: critical parameters and experimental process optimization approaches. Eur J Pharm Sci 127:300–318
25. Coimbra PPS, Cardoso FDSN, Gonçalves ÉCBDA (2021) Spray-drying wall materials: relationship with bioactive compounds. Crit Rev Food Sci Nutr 61:2809–2826

26. Jayasundera M, Adhikari B, Adhikari R et al (2011) The effect of protein types and low molecular weight surfactants on spray drying of sugar-rich foods. Food Hydrocolloids 25:459–469
27. Rahim MA, Imran M, Ambreen S et al (2023) Stabilization of the antioxidant properties in spray-dried microcapsules of fish and chia oil blends. ACS Omega 8:35183–35192
28. Swarbrick J (2013) Homogenization and homogenizers. In: Encyclopedia of pharmaceutical science and technology. 4th edn. CRC Press, Boca Raton, 1848–1854
29. Kang M, Park SY, Yun S-M et al (2024) Effects of sequential combination of cryogenic immersion freezing and ultra-cold frozen storage on the quality of Korean white kimchi. Innov Food Sci Emerging Technol 93:103642

Check for updates

Chapter 5

Microencapsulation of Bioactive Compounds by Spray-Chilling Technique

Fernanda Ramalho Procopio, Marluci Palazzolli da Silva-Padilha and Miriam Dupas Hubinger

Abstract

The microencapsulation process through the spray-chilling technique allows the incorporation of bioactive compounds into lipid matrices, forming solid particles at room temperature. The absence of solvent and the possibility of industrial scaling-up represent the main advantages of this technique. Due to the polymorphic changes and crystallization rate of lipid materials, a previous study of thermal behavior and compatibility between lipid matrices is necessary. This protocol presents the process parameters, equipment adaptations, and main characteristics of lipid carriers for obtaining microparticles with the best physical characteristics.

Keywords Solid lipid microparticles, Crystallization, Atomization

1 Introduction

For encapsulation purposes, a range of physical, chemical, and physicochemical methods can be employed. The selection is contingent upon the desired particle size, characteristics of the active ingredient, intended release mechanism, production scale, and cost. Among these methods, spray-chilling emerges as a versatile option due to its scalability and the absence of solvent during the process [1]. In fact, the same spray-drying equipment can be easily adapted for use as spray-chilling. The key difference lies in the use of lipid materials as carrier agents, and the equipment's ideal operation at a refrigeration temperature (< 10°C).

The main characteristics required for lipid materials used in the spray-chilling process are good crystallization properties, low cost, and compatibility with the active ingredient. These parameters are essential to ensure better particle flow and handling. In addition to better incorporate the active ingredient, empty spaces between the fatty acid chains and the crystalline lipid structure must be created [2].

Ana Silvia Prata and Louise Emy Kurozawa (eds.), *Biopolymers*, Methods and Protocols in Food Science,
https://doi.org/10.1007/978-1-0716-5230-5_5,

These spaces can be increased using lipid matrices with varying fatty acid composition or mixing low- and high-melting-point fats. The process begins with the direct incorporation of the active ingredient into the molten lipid carriers or the formation of a pre-emulsion. The mixture is then carefully pumped to an atomizing nozzle, which is heated above the melting temperature. This control is crucial to prevent premature crystallization and system blockage [2, 3]. Subsequently, the lipid mixture is atomized in a cold chamber, operating below the melting point. This controlled heat exchange facilitates the solidification of the material, giving rise to spherical matrix-type solid lipid microparticles (SLM) that are efficiently collected by the aspiration system. The resulting powder is easy to handle, transport, distribute, dose, and store.

Encapsulation by spray-chilling is a viable option for hydrophobic or hydrophilic active ingredientshydrophilic active ingredients. However, a prior compatibility study between the lipid matrices is recommended for lipophilic active ingredients such as oleoresins [3]. If this compatibility exists, the process gives rise to matrix-type particles, in which the active ingredient is thoroughly dispersed in the carrier material. As a result, part of the active compound may be deposited on the particle surface, reducing encapsulation efficiency and causing premature expulsion [4]. In some cases, mixing liquid and solid lipids can mitigate this effect. The presence of unsaturated and saturated fatty acid chains causes imperfections in the crystalline lipid matrix, favoring the accommodation of the lipophilic active ingredient [5]. Similarly, understanding the incorporation of hydrophilic active ingredients into the carrier agents is a significant step for a good encapsulation efficiency. As the matrix-type particles have the active ingredient dispersed in the carrier material, part of it can settle on the surface [1, 5]. Also the polymorphic rearrangements of lipids can promote a premature expulsion of the active ingredient. Hence, a thorough study of the polymorphic behavior of the lipid matrix is essential.

The polymorphism relates to the spatial organization of fatty acid chains during lipid crystallization [6]. Transitions from a less stable polymorphic form to one with greater thermodynamic stability can influence active retention. The matrix selected as an encapsulating agent must present a crystallization behavior compatible with the cooling process promoted by spray-chilling to guarantee a good crystalline structure capable of retaining the filling material [7, 8].

The objective of this protocol is to provide the necessary information on the preparation, operation, and characterization of solid lipid microparticles obtained by spray-chilling.

2 Materials and Equipment

2.1 Materials

- **Lipid carriers** (*see* **Note 1**): Palm fat, fully hydrogenated soybean oil, stearic acid.
- **Core materials** (*see* **Notes 2 and 3**): Oleoresins, omega-3-rich oils, carotenoids, vitamins, phenolic acids, and probiotics.

2.2 Equipments and Accessories

- Analytical balance.
- Beakers (50, 100, and 250 mL).
- Magnetic heating plate and magnetic bar.
- Microwave or water bath.
- Peristaltic pump.
- Thermometer.
- Graduated cylinder.
- Dry brush.
- Ultrasonic bath.
- Ultrasound probe.
- Ultra-Turrax homogenizer.

2.3 Equipment Requirements and Adaptations

One advantage of the spray-chilling technique is that it can be adapted from spray-dryer equipment. However, some adjustments are necessary to ensure good process reproducibility. First, the resistance that heats the drying air in the case of the spray dryer must remain off. Second, attaching a dehumidifier with refrigeration capacity to the equipment is interesting. This accessory allows the ambient air to be cooled before it enters the atomization chamber. Passing through the refrigeration system causes water condensation, releasing output air with a lower temperature and humidity than the input (ambient) air. This cooling air will promote the solidification of the lipid material and the formation of microparticles. Figure 1 shows the basic configuration of spray-chiller equipment.

1. **Feed system**: Some equipment has a spray-chilling accessory to be attached. The apparatus consists of a heating container for mixing the ingredients. In the absence of this accessory, an external adaptation to the spray-dryer equipment can be made. In this case, a magnetic heating plate keeps the mixture of active ingredients and carriers under heating and constant stirring (Fig. 1, item 1).

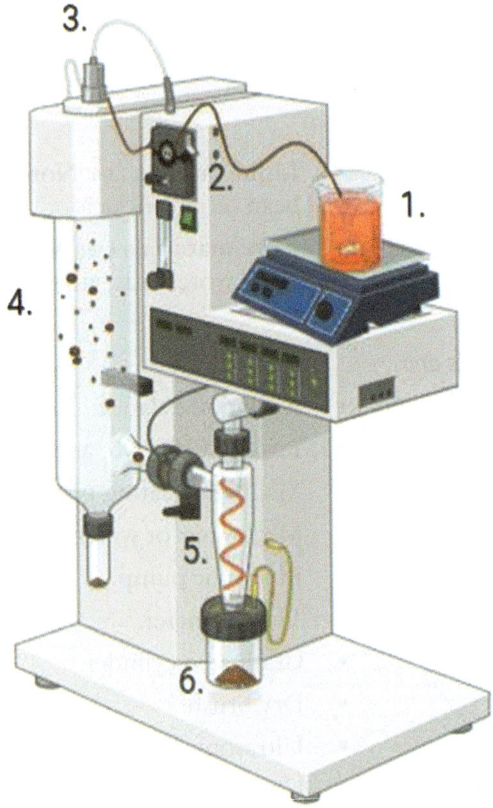

Fig. 1 Schematic of spray-chilling process. (1) Feed system, (2) pump, (3) atomization nozzle, (4) cold chamber, (5) cyclone separator, (6) powder storage. (Source: the author)

2. **Pump**: The pumping system must guide the mixture of active ingredients and carriers to the atomizing nozzle supply opening. Again, it can be done through the internal accessory or by attaching an external peristaltic pump (Fig. 1, item 2). The silicone hose must resist high temperatures, with a preferred external diameter of 14 mm.
3. **Atomizing nozzle**: The choice of atomizing nozzle size depends on the lipid matrix characteristics and the desired particle size. Double fluid atomizer nozzles of 0.7 mm [8–10], 1.2 mm [11–14], 2 mm [15], and 3 mm [16, 17] are suitable for the spray chilling process.
4. **Cold chamber**: The atomization chamber can be cooled using cold dehumidified air or liquid nitrogen. Typically, a cooling air flow of 35,000 L/h is used [9, 16, 17]. Similar to the spray-chilling process, the residence time of the sprayed droplets inside the cold chamber is short [4]. Therefore, the lower

the internal temperature, the more efficient the crystallization of lipids through heat transfer will be.

5. **Cyclone separator**: The separation cyclone captures the crystallized particles. Ideally, it should be coupled to a filtration container to prevent mist formation in the environment. The suction intensity (ideally 667–1052 L/h) can be adjusted to increase collection efficiency and prevent microparticles from escaping.
6. **Powder storage**: It is the compartment where the powder will be collected. However, higher process yields are achieved by also recovering particles adhered to the cooling chamber and cyclone.

3 Methods

3.1 Characterization of Lipid Carriers

3.1.1 Compatibility (see Note 4)

1. Weigh the components separately, exploring different ratios of hard fat and vegetable oil (e.g., 30:70, 50:50, and 70:30).
   Example: Place 70 g of hard fat (such as waxes, stearin, etc.) and 30 g of vegetable oil in separate beakers.
2. Heat each component in a microwave at medium power and for short intervals until completely melted. A water bath can be used, but care must be taken to avoid penetration of water vapor.
3. Use a thermometer to check that the temperature is at least 10°C above the melting point of the highest melting component.
4. Mix the melted lipids thoroughly with a glass rod or magnetic stirrer until homogeneous.
5. Allow the mixture to cool at room temperature and check that no phase separation occurs (no oil droplets or exudation on the surface).

Regarding hydrophilic bioactive compounds, some strategies can improve the compatibility between lipid carrier materials and the active ingredients (*see* **Notes 5 and 6**). In this case, it is suggested to produce a water-in-oil (W/O) emulsion [16]:

1. Heat the lipid phase (approximately 66 g) to about 10°C above the melting point of the hard fat. Stir the lipid phase using a magnetic stirrer at 500 rpm for 4 min.

2. Weigh approximately 30 g of the aqueous phase containing the bioactive compound (e.g., plant extract). Heat the aqueous phase to the same temperature range (±5°C) as the lipid phase to avoid thermal shock.
3. Dissolve polyglycerol polyricinoleate (PGPR, 4% w/w) in the melted fat until completely homogeneous before adding the aqueous phase.
4. Add the aqueous phase into the lipid phase under magnetic stirring at 800 rpm for 4 min. Carefully control the temperature of the mixture to prevent lipid crystallization, maintaining it at 10°C above the melting point.
5. If possible, subject the W/O emulsion to ultrasound treatment at 200 W for 4 min. This step can improve stability by reducing the droplet size of the dispersed phase.
6. Transfer an aliquot of the emulsion into a graduated cylinder and visually evaluate (no phase separation) its appearance for 10 min. Alternatively, perform a centrifugation test by placing samples in tubes at low rotational speed to accelerate phase separation.

3.1.2 Thermal Behavior (Differential Scanning Calorimetry [DSC])

The formation of stable particles, which have low adhesion and, therefore, greater yield in microencapsulation by spray-chilling, requires a lipid matrix with a melting point higher than 45°C [18]. To this end, an analysis of the thermal behavior of lipid mixtures is recommended. Thermal events can be visualized using the differential scanning calorimetry technique. Heat the mixtures until completely melted.

1. With the use of aluminum pans, carefully place approximately 5–10 mg of each formulation, ensuring the bottom is filled and avoiding any residue on the wall.
2. After closing, insert the pans into the location indicated on the equipment and carry out heating-cooling runs.
3. Heat the sample from room temperature to 80–85°C at a rate of 10°C/min in order to erase the thermal history.
4. Cool the sample to –20°C at a controlled rate of 10°C/min to allow visualization of the crystallization profile.
5. Reheat the sample to the desired temperature range at 10°C/min to identify endothermic melting events.

3.1.3 Crystalline Structure (X-Ray Diffraction Analysis [XRD])

Usually, the parameters used are Cu-Kα radiation (k = 1.5418 Å, 40 kV voltage, and 30 mA current), measured with a 2θ angle interval of 5°–40° at a scouring rate of 0.02°/s [3, 16, 19]. The polymorphic forms are identified based on characteristic short spacings (*see* **Note** 7).

3.2 Dispersing or Mixing the Active Ingredient into Lipid Carriers

3.2.1 Lipophilic Compounds

The mixture of lipophilic active ingredients can be made directly with molten carriers [3, 19] by magnetic stirring or using a pre-emulsion [20].

(a) Direct incorporation by magnetic stirring

1. Heat the lipid matrix above its melting point, keeping it under magnetic stirring for 20–30 min.
2. Ensure the carrier material is completely melted before adding the active ingredient.
3. Add the oily active ingredient carefully while stirring.

(b) Pre-emulsion

The formation of pre-emulsions that use some type of surfactant (polysorbate, polyglycerol polyricinoleate, soy lecithin) can be advantageous, guaranteeing greater system stability. This method can better incorporate interest compounds that have lower affinity for the lipid matrix or are difficult to solubilize [20]. For example:

1. Previously, mix the active ingredient with a low melting point lipid carrier (e.g., sunflower oil) and keep in an ultrasonic bath at 60 W and 25 ± 1°C for 10 min.
2. Add the surfactant (e.g., 1 g of soy lecithin) to the oily mixture and dropwise into an aqueous polymeric solution (protein, starch) under Ultra-Turrax homogenizer (10,000–20,000 rpm) for 3 min.
3. Add the final mixture to the melted (approximately 65°C) carrier material with a high melting point (e.g., fully hydrogenated palm oil) and homogenize for 3 min.

3.2.2 Hydrophilic Compounds

There are two main approaches for dispersing hydrophilic materials into melted lipid carriers: direct incorporation and hot emulsification.

(a) Direct incorporation [21]:

1. Heat until the lipid matrix is completely melted for 30 min.
2. Add the bioactive compounds in powder form.
3. Homogenize the mixture at 9000 rpm for 90 s, maintaining heat to prevent premature fat crystallization.

(b) Hot emulsification [8, 17]:

1. Conduct preliminary tests to determine the process parameters (emulsifier concentration, phase ratio, homogenization

speed and time, etc.) for forming water-in-oil (W/O) emulsions. The addition of emulsifiers in W/O emulsions, such as polyglycerol polyricinoleate (PGPR), is a common procedure to improve kinetics stability.

2. Solubilize the hydrophilic molecules in warm distilled water, maintaining the temperature according to the melting point of the lipid matrix.
3. Add the aqueous mixture containing the surfactant to the fat blend under constant stirring on a magnetic stirrer at 750 rpm for 4 min.
4. Homogenize in ultrasound for 4 min at 200 W power and 20 kHz to ensure greater stability.

4 Operational Spray-Chilling Proceedings

4.1 Equipment Preparation

1. Verify if all components, particularly the atomizing nozzle, are clean and in good condition, with no remnants of other materials present.
2. Connect the equipment according to the manufacturer's recommendations.
3. Perform experimental tests to optimize the processing parameters.

4.2 Setting Operational Parameters

1. Set the air flow according to the manufacturer's recommendation.
2. Set the inlet temperature to the atomization chamber (depending on the lipid matrix, temperatures can vary from 7 to 13°C). Afterward, turn on the aspirator (maximum). This step may be time-consuming depending on the equipment used, so the preparation of the feeding matrix can be conducted simultaneously.
3. Turn on the heating system to maintain control over the atomizer nozzle temperature. If there is no atomizer nozzle heating system, an alternative is to heat it in an oven. This process will prevent premature crystallization of the lipid matrix during feeding.
4. Adjust the feed flow following the manufacturer's recommendations and preliminary tests.

4.3 Operating the Spray-Chiller

1. Use thermal gloves to handle the container containing the lipid matrix, as well as the silicone hose and atomizing nozzle. It is also important to wear protective glasses in case the silicone hose or atomizer nozzle becomes clogged.

2. Place the heated auxiliary accessories onto the equipment just before starting the feeding process. Turn on the magnetic heating plate to keep the mixture of active ingredients and carriers heated and continuously stirred (*see* **Note 8**).
3. Connect the feed pump to introduce the lipid matrix into the equipment (*see* **Note 9**).
4. Check during feeding whether the material is being atomized. If not, there may be an issue with the compressed air flow, and the operation should be halted until the problem is resolved (*see* **Note 10**).
5. Turn off the feed pump once all the lipid material has been pumped into the equipment, followed by shutting down the air flow pressure and temperature control system of the device.
6. Carefully detach the separator cyclone from the main equipment unit and recover the solid lipid microparticles into a suitable container or collection vessel (*see* **Note 11**).

4.4 Cleaning

Clean the dismantled parts thoroughly to remove any remaining particles or residue. Utilizing heated tap water, a microwave, or a water bath can facilitate this cleaning process, effectively removing any remaining lipid material.

4.5 Microparticles Characterization (see Note 12)

- The particle should be evaluated for size as described in the chapter "Characterization."
- Crystallinity (DRX): Use Cu-Kα source to generate X-rays of $\lambda = 1.54056$ Å. Measure the diffraction in the 2θ range, from 5° to 40°, at a scouring rate of 0.02°/s.
- Thermal behavior (DSC): Weigh approximately 5 mg into hermetic aluminum pans. Heat the sample to a temperature above the melting point of the carriers (maintaining for 5 min) at a rate of 10°C/min. Cool the sample to −20 °C at a cooling rate of 10°C/min. Reheat the sample under the same conditions as those used for the first heat curve.
- Stability: Analyze particle size, thermal behavior, and crystallinity over time (30–120 days) under controlled storage conditions (25°C, 50% relative humidity) (*see* **Note 13**). Chromatographic and spectrophotometric techniques can also be used to evaluate the core material stability [3, 22]. For this:
    - Dissolve the particles in a solvent (such as water or ethanol) and extract the active compound. Heating and an ultrasonic bath can also be used to help break up the particles [23].

5 Notes

1. Vegetable oils, fats, waxes, and free fatty acids (stearic, palmitic, oleic acid) are the main matrices used as carrier materials. They can be used individually or in a mixture, and must present a melting temperature above 45°C [5, 18]. Palm fat [24, 25], soybean oil, and commercial vegetable fats [25, 26] appear as the most used carrier materials of low/medium melting points. Fully hydrogenated vegetable fats stand out among the lipid matrices with the highest melting points [8, 27].
2. Lipophilic bioactive compounds such as carotenoids [11, 23, 26], oleoresins [3, 9], and omega-3-rich oils [28] have been encapsulated by spray-chilling- due to their greater affinity for the lipid matrix.
3. Over the years, several hydrophilic bioactive molecules have been microencapsulated using spray-chilling, including vitamins [22], probiotics [12, 29], phenolic acids [8], and more. In relation to plant extracts, they contain molecules with different polarities, interacting in different ways with the lipid matrix during encapsulation by spray-chilling.
4. Mixing different lipid matrices can enhance the incorporation of lipophilic active ingredients. However, it is crucial to ensure that these mixtures are compatible.
5. Due to low compatibility between the carrier material and the hydrophilic molecule, some strategies can improve the encapsulation efficiency. These include selecting and testing the lipid matrix and the bioactive compound, producing oil-in-water emulsions using emulsifiers and checking the kinetic stability or performing the direct dispersion of powder containing the bioactive compound.
6. Add antioxidants into solid lipid microparticles formulation to improve the preservation of hydrophilic bioactive compounds and probably mitigate the lipid oxidation during storage.
7. The specific properties of oil and fat crystallization may be related to polymorphism or spatial organization between fatty acid chains. Polymorphism is the ability of a chemical compound to form different crystalline structures, with the α, β', and β forms typical for oils and fats following this same order of increasing stability [29]. Generally, a β' characteristic is preferred for spray-chilling application.
8. Selecting appropriate homogenization parameters and controlling the temperature for melting lipid carriers and homogenization using a water bath heater are crucial to

avoid premature crystallization of the lipid matrix, as well as to enhance encapsulation efficiency.

9. The active ingredient must be added immediately before starting the atomization process to avoid compound degradation.
10. Diameters smaller than 14 mm may cause clogging during the process. It is worth mentioning that when using lipid mixtures with a high melting point (>70°C), a "jacket" system is recommended to keep the hose heated, preventing crystallization and clogging.
11. A dry brush can help recover the dry particles at the end of the spray-chilling process.
12. The type of intended application should be considered when evaluating X-ray diffraction (DRX), dynamic light scattering (DLS), and differential scanning calorimetry (DSC) under storage conditions, including time (30–180 days), temperature (6–45°C), and relative humidity (40%–70%).
13. Differential scanning calorimetry (DSC) is used to evaluate changes in the crystallinity and thermal behavior of lipid materials [3, 19, 30]. These measurements help investigate the reorganization of lipid chains and polymorphic transitions. The occurrence of polymorphic changes over time, as evidenced by new or suppressed peaks in the thermal profile, can influence the release of the active ingredient. Similarly, the increase in average particle size, signaling the formation of agglomerates, may not align with desired outcomes [31]. In such cases, a change in the encapsulating lipid matrix could be warranted [31].

References

1. Bertoni B, Albertini B, Passerini N (2019) Spray congealing: an emerging technology to prepare solid dispersions with enhanced oral bioavailability of poorly water soluble drugs. Molecules 24:3–5. https://doi.org/10.3390/molecules24193471
2. Müller RH, Radtke M, Wissing SA (2002) Nanostructured lipid matrices for improved microencapsulation of drugs. Int J Pharm 242:121–128. https://doi.org/10.1016/S0378-5173(02)00180-1
3. Oriani VB, Alvim ID, Consoli L, Molina G, Pastore GM, Hubinger MD (2016) Solid lipid microparticles produced by spray chilling technique to deliver ginger oleoresin: structure and compound retention. Food Res Int 80:41–49. https://doi.org/10.1016/j.foodres.2015.12.015
4. Figueiredo JdeA, Silva CRdeP, Souza Oliveira MF, Norcino LB, Campelo PH, Botrel DA, Borges SV (2022) Microencapsulation by spray chilling in the food industry: opportunities, challenges, and innovations. Trends Food Sci Technol 120:274–287. https://doi.org/10.1016/j.tifs.2021.12.026
5. Alvim ID, Souza FDS, Koury IP, Jurt T, Dantas BFH (2013) Use of the spray chilling method to deliver hydrophobic components: physical characterization of microparticles. Ciência e Tecnologia de Alimentos 33:34–39. https://doi.org/10.1590/S0101-20612013000500006
6. McClements DJ (2012) Crystals and crystallization in oil-in-water emulsions: implications for emulsion-based delivery systems. Adv Colloid Interface Sci 174:1–30. https://doi.org/10.1016/j.cis.2012.03.002

7. da Silva Santos V, Badan Ribeiro AP, Andrade Santana MH (2019) Solid lipid nanoparticles as carriers for lipophilic compounds for applications in foods. Food Res Int 122:610–626. https://doi.org/10.1016/J.FOODRES.2019.01.032
8. Consoli L, Grimaldi R, Sartori T, Menegalli FC, Hubinger MD (2016) Gallic acid microparticles produced by spray chilling technique: production and characterization. LWT - Food Sci Technol 65:79–87. https://doi.org/10.1016/j.lwt.2015.07.052
9. Procopio FR, Klettenhammer S, Ferrentino G, Scampicchio M, do Amaral Sobral PJ, Hubinger MD (2023) Comparative study of cinnamon and paprika oleoresins encapsulated by spray chilling and particles from gas saturated solutions techniques: evaluation of physical characteristics and oleoresins release in food simulated media. Food Bioprocess Technol 16:2147–2158. https://doi.org/10.1007/s11947-023-03058-5
10. Rebellato AP, de Moraes PP, Silva JGS, Alvim ID, Lima Pallone JA, Steel CJ (2024) Ferrous sulfate microparticles obtained by spray chilling: characterization, stability and *in vitro* digestion simulation. J Food Sci Technol 61:97–105. https://doi.org/10.1007/s13197-023-05820-1
11. de PD, Santos F, do L, Siqueira V, Tadini CC, Favaro-Trindade CS (2023) Characterization of cassava starch extruded sheets incorporated with tucumã oil microparticles. Processes 11:876. https://doi.org/10.3390/pr11030876
12. Silva MP, Tulini FL, Matos FE, Oliveira MG, Thomazini M, Fávaro-Trindade CS (2018) Application of spray chilling and electrostatic interaction to produce lipid microparticles loaded with probiotics as an alternative to improve resistance under stress conditions. Food Hydrocoll 83:109–117. https://doi.org/10.1016/j.foodhyd.2018.05.001
13. Gomes Sá SH, Chalella Mazzocato M, Saliba ASMC, Alencar SM, Sílvia Favaro-Trindade C (2023) Evaluation of the release, stability and antioxidant activity of Brazilian red propolis extract encapsulated by spray-drying, spray-chilling and using the combination of both techniques. Food Res Int 164:112423. https://doi.org/10.1016/j.foodres.2022.112423
14. Tulini FL, Souza VB, Echalar-Barrientos MA, Thomazini M, Pallone EMJA, Favaro-Trindade CS (2016) Development of solid lipid microparticles loaded with a proanthocyanidin-rich cinnamon extract (*Cinnamomum zeylanicum*): potential for increasing antioxidant content in functional foods for diabetic population. Food Res Int 85:10–18. https://doi.org/10.1016/j.foodres.2016.04.006
15. dos JD, Carvalho S, Oriani VB, de Oliveira GM, Hubinger MD (2019) Characterization of ascorbic acid microencapsulated by the spray chilling technique using palm oil and fully hydrogenated palm oil. LWT 101:306–314. https://doi.org/10.1016/j.lwt.2018.11.043
16. Oliveira MFS, Figueiredo JdeA, Norcino LB, Botrel DA, Borges SV (2023) Potential use of red hibiscus flower extract for the production of spray-chilled microparticles: characterization, stability, and bioaccessibility *in vitro* of anthocyanins. Food Res Int 174:113570. https://doi.org/10.1016/j.foodres.2023.113570.
17. de Abreu Figueiredo J, Norcino LB, do Carmo EL, Campelo PH, Botrel DA, Borges SV, de Souza SM, de Oliveira CR (2023) Microstructured lipid microparticles containing anthocyanins: production, characterization, storage, and resistance to the gastrointestinal tract. Food Res Int 166:112611. https://doi.org/10.1016/j.foodres.2023.112611
18. Karami A, Abroumand Gholami A, Babaloo H, Farhadian N (2023) Encapsulation: spray chilling and cooling. In: Sefat F, Farzi G, Mozafari M (Eds.) Principles of biomaterials encapsulation. Elsevier, United States, vol 1 109–130 https://doi.org/10.1016/B978-0-323-85947-9.00003-0
19. Procopio FR, Oriani VB, Paulino BN, do Prado-Silva L, Pastore GM, Sant'Ana AS, Hubinger MD (2018) Solid lipid microparticles loaded with cinnamon oleoresin: characterization, stability and antimicrobial activity. Food Res Int 113:351–361. https://doi.org/10.1016/J.FOODRES.2018.07.026
20. Günel Z, Varhan E, Koç M, Topuz A, Sahin-Nadeem H (2021) Production of pungency-suppressed capsaicin microcapsules by spray chilling. Food Biosci 40:100918. https://doi.org/10.1016/j.fbio.2021.100918.
21. Cutrim CS, Alvim ID, Cortez MAS (2019) Microencapsulation of green tea polyphenols by ionic gelation and spray chilling methods. J Food Sci Technol 56:3562–3563. https://doi.org/10.1007/s13197-019-03908-1
22. dos JD, Carvalho S, Oriani VB, de Oliveira GM, Hubinger MD (2019) Characterization of ascorbic acid microencapsulated by the spray chilling technique using palm oil and fully hydrogenated palm oil. LWT 101:306–314. https://doi.org/10.1016/j.lwt.2018.11.043.
23. de Lima PM, Dacanal GC, Pinho LS, de Sá SHG, Thomazini M, Favaro-Trindade CS (2022) Combination of spray-chilling and spray-drying techniques to protect carotenoid-rich extracts from pumpkin (*Cucurbita moschata*) Byproducts, aiming at the production of a powdered natural food Dye. Molecules 27:7530. https://doi.org/10.3390/molecules27217530
24. Günel Z, Varhan E, Koç M, Topuz A, Sahin-Nadeem H (2021) Production of

pungency-suppressed capsaicin microcapsules by spray chilling. Food Biosci 40:100918. https://doi.org/10.1016/j.fbio.2021.100918
25. Queiroz de Oliveira W, Angélica Neri Numa I, Alvim ID, Azeredo HMC, Santos LB, Borsoi FT, de Araújo FF, Sawaya ACHF, do Nascimento GC, Clerici MTPS, do Sacramento CK, Maria Pastore G (2024) Multilayer microparticles for programmed sequential release of phenolic compounds from *Eugenia stipitata*: stability and bioavailability. Food Chem 443:138579. https://doi.org/10.1016/j.foodchem.2024.138579
26. Pinho LS, Lima PM, Fang F, Cooperstone JL, Favaro-Trindade CS, Campanella OH (2024) Effect of extrusion process conditions on extrudates enriched with carotenoids encapsulated by different methods using gum arabic and vegetable fat as carriers. Int J Biol Macromol 267:131200. https://doi.org/10.1016/j.ijbiomac.2024.131200
27. Gamboa OD, Gonçalves LG, Grosso CF (2011) Microencapsulation of tocopherols in lipid matrix by spray chilling method. Procedia Food Sci 1:1732–1739. https://doi.org/10.1016/j.profoo.2011.09.255
28. Fadini AL, Alvim ID, Carazzato CA, de KB, Paganotti F, de AMR, Miguel O, Rodrigues RAF (2021) Microparticles loaded with fish oil: stability studies, food application and sensory evaluation. J Microencapsul 38:365–380. https://doi.org/10.1080/02652048.2021.1948622
29. Arslan-Tontul S, Erbas M (2017) Single and double layered microencapsulation of probiotics by spray drying and spray chilling. LWT Food Sci Technol 81:160–169. https://doi.org/10.1016/J.LWT.2017.03.060
30. Sato K, Ueno S (2011) Crystallization, transformation and microstructures of polymorphic fats in colloidal dispersion states. Curr Opin Colloid Interface Sci 16:384–390. https://doi.org/10.1016/j.cocis.2011.06.004
31. dos S. Carvalho JD, Oriani VB, Oliveira GM, Hubinger MD (2021) Solid lipid microparticles loaded with ascorbic acid: release kinetic profile during thermal stability. J Food Process Preserv 45:2–9. https://doi.org/10.1111/jfpp.15557

Check for updates

Chapter 6

Solvent Evaporation

Gabriel Pedroso De Lima Alexandre and Ana Silvia Prata

Abstract

Solvent evaporation is a widely applied encapsulation technique in the food, pharmaceutical, and cosmetic industries, owing to its versatility and ability to produce particles with controlled size and morphology. In this method, a polymer and core material are dissolved in a volatile organic solvent, which is then emulsified into an aqueous continuous phase. As the solvent diffuses and evaporates, the polymer solidifies, entrapping the core material within micro- or nanoparticles. Depending on the solubility of the core, single oil-in-water (O/W), double water-in-oil-in-water (W/O/W), or polyol-oil-polyol (P/O/P) emulsions may be adopted. This protocol describes step-by-step methodologies for encapsulation of hydrophilic and hydrophobic core materials using the solvent evaporation technique.

Keywords Organic solvent, Colligative properties, Hydrophobic drugs, Hydrophilic drugs, Double emulsion

1 Introduction

In solvent evaporation encapsulation, the core material and polymer are dissolved in an organic solvent. Then, this organic phase is emulsified into an aqueous phase containing a suitable emulsifier, which results in an O/W emulsion (droplet formation). Afterward, the organic solvent diffuses and evaporates from the droplets into the continuous phase, leading to the phase separation and polymer solidification. Once the organic solvent is evaporated altogether, the polymer migrates around the surface of the core material due to surface tension interactions, forming microcapsules. Figure 1 illustrates the overall process for producing hydrophobic drug-loaded microcapsules by solvent evaporation.

To control the size and morphology of the drug-loaded particles, parameters, such as evaporation and stirring rates, are adjusted, but the viscosity of the emulsion will also affect. Moreover, it is suitable to use a solvent with high volatility and low boiling point to avoid high temperature ranges [1–3].

Ana Silvia Prata and Louise Emy Kurozawa (eds.), *Biopolymers*, Methods and Protocols in Food Science, https://doi.org/10.1007/978-1-0716-5230-5_6,

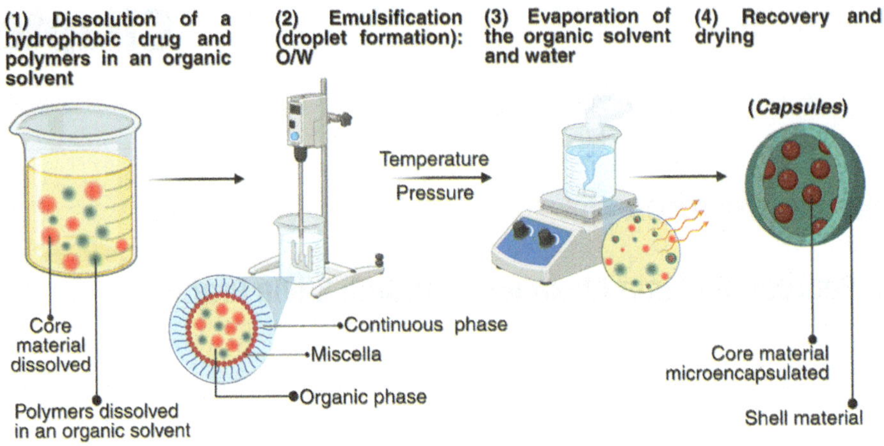

Fig. 1 Step-by-step process for microencapsulation of hydrophobic drug-loaded core material

The mean diameter of particles ranges from 10 to 300 μm, depending on polymer concentration, molecular mass of polymer, organic phase viscosity, and stirring rate [4, 5].

In the case of hydrophilic core material encapsulation, the primary W/O emulsion containing the water-soluble compound is added to another aqueous solution with an emulsifier to produce a water-in-oil-in-water (W/O/W) double emulsion [1]. Double emulsion was applied to iohexol [6], insulin [7, 8], proteins [9–11], and vaccines [12, 13]. Some compounds, such as flavonoids, are poorly water-soluble and have low solubility in organic solvents; in these cases, a polyol-in-oil-in-polyol (P/O/P) emulsion can be adopted [3].

The most common organic solvents are chloroform, dichloromethane, ethyl acetate, and ethyl [1], and the more used shell materials are biodegradable synthetic biopolymers, such as polyhydroxybutyrate (PHB), polycaprolactone (PCL), poly(lactic acid) (PLA), or poly(lactide-co-glycolide) (PLGA) [2].

2 Materials

2.1 Hydrophobic Core Material

- Curcumin (core material) (*see* **Note 1** for additional hydrophobic core materials).
- Dichloromethane (DCM) (organic phase).
- Poly(D,L-lactide-co-glycolide) (PLGA) (as shell material) (*see* **Notes 2 and 3**).
- Poly(vinyl alcohol) (PVA) (partially hydrolyzed: 87%–90%; MW ≅ 30–70 kDa; as emulsifier).

2.2 Hydrophilic Core Material

- Ascorbic acid (core material) (*see* **Note 4** for additional hydrophilic core materials).
- Paraffin (organic solvent).
- Hexane (organic solvent).
- Ethyl cellulose (45 cP) (as plasticizer).
- Absolute ethanol (continuous phase).

2.3 Equipment

- Centrifuge.
- Cooling system/condenser.
- Freeze-dryer.
- Glass syringes.
- Heating bath.
- Magnetic stirrer.
- Rotary evaporator.
- Rotary-evaporator flask.
- Sonicator.
- Ultrasonic probe.
- Ultrasonic processor.
- UltraTurrax.
- Vacuum pump + controller.

3 Methods

3.1 Encapsulation of Hydrophobic Core Material

1. Prepare carefully 0.5% (m/v) PVA (aqueous phase).
2. Place 10 mL of 0.5% (m/v) PVA into a beaker under constant stirring at 800 rpm.
3. Dissolve 250 mg of PLGA and 25 mg of curcumin in 5 mL of DCM until reach a clear solution (organic phase).
4. Transfer the organic solution (Step 3) into a glass syringe.
5. With the aqueous phase stirring at 800 rpm, add the organic solution (Step 3) dropwise into the 10 mL PVA. After finishing the addition, continue stirring for 10 s. This produces an O/W emulsion.
6. Immerse the beaker in an ice bath at once. Insert the 3-mm diameter ultrasonic probe into the O/W emulsion and avoid touching the beaker wall and bottom (*see* **Note 5**).
7. Sonicate the O/W emulsion for 3 min at 13 W and 40% amplitude. Maintain the sample cooled on ice during sonication to avoid heating and curcumin degradation.

8. Pour the sonicated dispersion into 40 mL of 0.5% PVA (pre-stirring at 800 rpm). Stir the combined dispersion at 800 rpm for 1 min.
9. Transfer dispersion to the rotary-evaporator flask. Evaporate DCM at 35°C under reduced pressure (approximately 50 mbar) until solvent is removed and the dispersion concentrated to a final volume ranging 10–15 mL.
10. Centrifugate the particles (Step 9) at 42,000 × *g* for 20 min at 20°C (*see* **Note 6**).
11. Throw away the supernatant.
12. Resuspend the particles in water and freeze at −80°C in appropriate vessels for freeze-dryers.
13. Run freeze-drying in Auto Mode (collector temperature −40°C, pressure ≤ 0.133 mbar) for about 24 h.
14. Store the freeze-dried curcumin-loaded particles in airtight packaging and protected from light.

3.2 Hydrophilic Core Material

1. Prepare the continuous phase mixing 210 mL liquid paraffin and hexane at 2:1 ratio (paraffin:hexane). Put it aside (*see* **Note 7**).
2. Dissolve 2.0 g of ethyl cellulose (45 cP) in 10 mL absolute ethanol using a water bath at 40°C (polymer mixture) (*see* **Note 8**).
3. Add the required amount of ascorbic acid to the polymer mixture above.
4. Emulsify using a sonicator with a frequency of 70 kHz to obtain a stable primary W_1/O emulsion.
5. Transfer the W_1/O emulsion into the continuous phase (Step 1) meticulously.
6. Emulsify the previous mixture to form W_1/O/W_2 emulsion (double emulsion).
7. Evaporate the double emulsion at 28–55°C and continuous stirring at 1000 rpm.
8. After the complete evaporation of alcohol, collect the microcapsules by filtration (*see* **Note 9**).
9. Wash the particles with hexane to remove residual paraffin and solvent. Repeat until there is no residual solvent.
10. Freeze-dry the ascorbic-loaded particles.
11. Store the freeze-dried ascorbic-acid-loaded particles in airtight packaging and protected from light.

4 Notes

1. Curcumin + poly(D,L-lactic acid)-glycerol [14]; resveratrol in stearic acid solid matrix + Phospholipon® 90G:Pluronic® F-68 mixture [15]; itraconazole + mesoporous silica and nifedipine + mesoporous silica [16]; annatto extract + poly(3-hydroxybutyrate-co-3-hydroxyvalerate) [17]; lycopene + polycaprolactone [18].
2. Polycaprolactone (PCL) and Eudragit® RL PO can also be used as shell materials. Similar to PLGA, PCL provides high encapsulation efficiency and drug loading, while Eudragit® RL PO is a promising carrier for oral delivery owing to its mucoadhesive properties and rapid drug release, which may enhance bioavailability.
3. β-cyclodextrin can be used as a shell material. See Ref. [19] for further details.
4. Nicotinamide + mesoporous silica [16]; linezolid + polycaprolactone (PCL) [20]; swine insulin + poly(lactide-co-glycolide) [21]; 5-aminosalicylic acid + Eudragit® S100 [22]; donepezil hydrochloride + poly(lactic-co-glycolic acid) [23].
5. UltraTurrax can be employed to provide high-speed homogenization. By employing a homogenizer, Lemoine and Préat [10] specify levels of rotational speed for both O/W and $W_1/O/W_2$.
6. Some studies rinse the particles with distilled water many times (Step 10) before following the next step. In addition, centrifugation settings are conducted at 8000 × *g* for 8 min in other studies [24].
7. To obtain the required volume at a 2:1 paraffin:hexane ratio, use 140 mL of paraffin and 70 mL of hexane.

   Mathematically,

$$\begin{cases} \text{Paraffin} = 2X \\ \text{Hexane} = X \\ \text{Paraffin} + \text{Hexane} = 210\ \text{mL} = 2X + X \\ \therefore X = 70\ \text{mL} \end{cases}$$

8. Adding 1 mL of triethyl citrate to the polymer solution (Step 1) provides a slower and more controlled release of the core material by plasticizing ethyl cellulose polymer chains, forming a harder and more uniform wall and reducing the wall porosity [25].
9. Confirm particle size beforehand to select an appropriate filter paper or Büchner funnel, using either optical microscopy or sieving on a mechanical shaker with a standard sieve for 5 min [25].

References

1. Li M, Rouaud O, Poncelet D (2008) Microencapsulation by solvent evaporation: state of the art for process engineering approaches. Int J Pharm 363(1-2):26–39. https://doi.org/10.1016/j.ijpharm.2008.07.018
2. Gundloori RV, Singam A, Killi N (2019) Chapter 19 -Nanobased intravenous and transdermal drug delivery systems. In: Mohapatra SS, Ranjan S, Dasgupta N, Mishra RK, Thomas S (Eds.) In Micro and Nano Technologies, Applications of Targeted Nano Drugs and Delivery Systems. Elsevier, 551–594 https://doi.org/10.1016/B978-0-12-814029-1.00019-3 ISBN 9780128140291
3. Lee D-H, Sim G-S, Kim J-H, Lee G-S, Pyo H-B, Lee B-C (2007) Preparation and characterization of quercetin-loaded polymethyl methacrylate microcapsules using a polyol-in-oil-in-polyol emulsion solvent evaporation method. J Pharm Pharmacol 59(12):1611–1620. https://doi.org/10.1211/jpp.59.12.0002
4. McGinity JW, O'Donnell PB (1997) Preparation of microspheres by the solvent evaporation technique. Adv Drug Deliv Rev 28(1):25–42. https://doi.org/10.1016/s0169-409x(97)00049-5
5. André-Abrant A, Taverdet J-L, Jay J (2001) Microencapsulation par Évaporation de Solvant. Eur Polym J 37(5):955-963, ISSN 0014-3057, https://doi.org/10.1016/S0014-3057(00)00197-X
6. Ogata Y, Kuroiwa T, Ichikawa S (2023) Facilitated encapsulation of a nonionic contrast agent for X-Ray computed tomography into lipid vesicles by the multiple emulsification-solvent evaporation method. Colloids Surf B: Biointerfaces 227:113360, ISSN 0927-7765, https://doi.org/10.1016/j.colsurfb.2023.113360
7. Singh M, Shirley B, Bajwa K, Samara E, Hora M, O'Hagan D (2001) Controlled release of recombinant insulin-like growth factor from a novel formulation of polylactide-co-glycolide microparticles. J Control Release 70 (1): 21–28, ISSN 0168-3659, https://doi.org/10.1016/S0168-3659(00)00313-8
8. Meinel L, Illi OE, Zapf J, Malfanti M, Peter Merkle H, Gander B (2001) Stabilizing insulin-like growth factor-I in poly(D,L-lactide-co-glycolide) microspheres.. J Control Release 70(1-2):193–202. https://doi.org/10.1016/s0168-3659(00)00352-7
9. Diwan M, Park TG (2001) Pegylation enhances protein stability during encapsulation in PLGA microspheres. J Control Release 73(2–3). https://doi.org/10.1016/S0168-3659(01)00292-9
10. Lemoine D, Préat V (1998) Polymeric nanoparticles as delivery system for influenza virus glycoproteins. J Control Release 54(1):15-27. https://doi.org/10.1016/S0168-3659(97)00241-1
11. Tapfumaneyi P, Imran M, Alavi SE, Mohammed Y (2023) Science of, and insights into, thermodynamic principles for dermal formulations. Drug Discov Today 28(4):103521. https://doi.org/10.1016/j.drudis.2023.103521
12. Little SR, Lynn DM, Puram SV, Langer R (2005) Formulation and characterization of poly (β amino ester) microparticles for genetic vaccine delivery. J Control Release 107(3):449-462 https://doi.org/10.1016/j.jconrel.2005.04.022
13. Sah H, Toddywala R, Chien YW (1995) Continuous release of proteins from biodegradable microcapsules and *in vivo* evaluation of their potential as a vaccine adjuvant. J Control Release 35(2–3):137-144. https://doi.org/10.1016/0168-3659(95)00029-8
14. Yoon I-S, Park J-H, Kang HJ, Choe JH, Goh MS, Kim D-D et al. (2015) Poly(D,L-lactic acid)-glycerol-based nanoparticles for curcumin delivery. Int J Pharm 488(1-2):70–77. https://doi.org/10.1016/j.ijpharm.2015.04.046
15. Pandita D, Kumar S, Poonia N, Lather V (2014) Solid lipid nanoparticles enhance oral bioavailability of resveratrol, a natural polyphenol. Food Res Int 62:1165–1174. https://doi.org/10.1016/j.foodres.2014.05.059
16. Budiman A, Aulifa DL (2021) Encapsulation of drug into mesoporous silica by solvent evaporation: a comparative study of drug characterization in mesoporous silica with various molecular weights. Heliyon 7(12):e08627. https://doi.org/10.1016/j.heliyon.2021.e08627
17. Teixeira Z, Durán N, Guterres SS (2008) Annatto polymeric microparticles: natural product encapsulation by the emulsion–solvent evaporation method. J Chem Educ 85(7):946–947. https://doi.org/10.1021/ed085
18. Li J, Bufalini C, Mottola S, Iannaco MC, Campardelli R, Marco ID (2026) Valorisation of tomato peel waste for lycopene encapsulation: optimization and comparison of two green techniques. J Supercrit Fluids 229:106803. https://doi.org/10.1016/j.supflu.2025.106803
19. Yallapu MM, Jaggi M, Chauhan SC (2010) β-Cyclodextrin-curcumin self-assembly enhances curcumin delivery in prostate cancer cells. Colloids Surf B: Biointerfaces 79(1):113–125. https://doi.org/10.1016/j.colsurfb.2010.03.039
20. Sharma Y, Mahar R, Chakraborty A, Nainwal N (2023) Optimizing the formulation variables for encapsulation of linezolid into polycaprolactone inhalable microspheres using double emulsion

solvent evaporation. Tuberculosis 143:102417. https://doi.org/10.1016/j.tube.2023.102417
21. Bao W, Zhou J, Luo J, Wu D (2006) PLGA microspheres with high drug loading and high encapsulation efficiency prepared by a novel solvent evaporation technique. J Microencapsul 23(5):471–479. https://doi.org/10.1080/02652040600687613
22. Vizcaya D, Tirado DF, Cabañas A, Serrano DR, Calvo L (2026) Preparation of 5-ASA-loaded Eudragit® S100 nanoparticles by emulsion-based methods: comparison between solvent evaporation and supercritical fluid extraction. J Supercrit Fluids 227: 106753. https://doi.org/10.1016/j.supflu.2025.106753
23. Kömür M, Kıyan HT, Öztürk AA (2025) Development of donepezil hydrochloride-loaded PLGA-based nanoparticles for Alzheimer's disease treatment. Sci Rep 15(1):13184. https://doi.org/10.1038/s41598-025-95792-3
24. Li X, Deng X, Yuan M, Xiong C, Huang Z, Zhang Y et al (1999) Investigation on process parameters involved in preparation of poly-DL-lactide-poly(ethylene glycol) microspheres containing *Leptospira interrogans* antigens. Int J Pharm 178(2):245-255. https://doi.org/10.1016/S0378-5173(98)00379-2
25. Uddin MS, Hawlader MNA, Zhu HJ (2001) Microencapsulation of ascorbic acid: effect of process variables on product characteristics. J Microencapsul 18(2):199209. https://doi.org/10.1080/02652040010000352

Check for updates

Chapter 7

Nanoencapsulation of Lipid Particles

Thaís J. Silva, Fernanda L. Lüdtke and Ana Paula B. Ribeiro

Abstract

Lipids have been used as raw materials for the preparation of nanostructures, mainly due to their ability to solubilize lipophilic bioactive compounds (BC). This chapter brings together information about lipid materials, emulsifiers, and BC, as well as low- and high-energy methods used in the preparation of lipid nanoparticles (LN). Important methods for characterizing LN in terms of zeta potential, morphology, crystallization behavior, and polymorphism are presented, seeking to support future studies.

Keywords Solid lipid nanoparticles, Nanostructured lipid carriers, High-pressure homogenization, Lipid nanoparticle formulation, Characterization methods

1 Introduction

Nanoparticles are defined as discrete entities with at least one dimension, generally smaller than 100 nm. This size characteristic, combined with their surface structure and chemical composition, gives nanoparticles unique application characteristics. A wide variety of nanoparticles can be applied in the food industry, particularly solid particles and carrier systems developed at the nanoscale [1].

Lipid nanoparticles (LNs) are among the most promising encapsulation technologies in the field of nanostructures. Compared to conventional encapsulation systems, lipid systems offer the advantage of being able to be produced on an industrial scale with natural ingredients, significantly differentiating the physicochemical properties of the LNs obtained, and the ability to retain compounds with widely varying solubilities. They are generally characterized by dimensions between 50 and 1000 nm, are physiologically tolerable, and are solid at room or body temperature [2, 3]. The great potential for application of LN refers to the enormous versatility in obtaining systems whose physical properties, such as dimensions, structure, charge, physical state, and preferred crystalline habits, can be controlled through the selection of specific lipid raw materials, associated with different production parameters [1].

Ana Silvia Prata and Louise Emy Kurozawa (eds.), *Biopolymers*, Methods and Protocols in Food Science,
https://doi.org/10.1007/978-1-0716-5230-5_7,

LNs are currently classified into two categories: solid lipid nanoparticles (SLN) and nanostructured lipid carriers (NLC). The production of SLN is based on the concept of replacing the aqueous core of emulsions with solid lipids. The first generation of SLN was produced solely with solid lipids and emulsifiers, and a second generation, the NLCs, was developed using solid, liquid, and emulsifier lipid systems. This second generation emerged with the aim of producing a lipid matrix with less crystallinity structure, capable of achieving better encapsulation efficiency and avoiding the release of incorporated compounds during storage [3].

SLNs are composed of high-melting-point lipids dispersed in an aqueous medium in concentrations between 0.1% and 30% (w/w) and stabilized by 0.5%–5% (w/w) of emulsifiers, while for NLCs the percentage of lipids in the formulation can exceed 95% (w/w). In both systems, the percentage of the active compound in relation to the total formulation is close to 5% (w/w) [1, 2]. Figure 1 presents a schematic representation of the LN structure.

LNs are produced by two main groups of methods, classified according to the amount of energy required in the process: (i) using high energy, which is provided by equipment such as high-pressure homogenizers, high-speed homogenization, and ultrasound and (ii) using low energy, which includes the use of controllable mixing processes under low agitation (usually using a magnetic stirrer), such

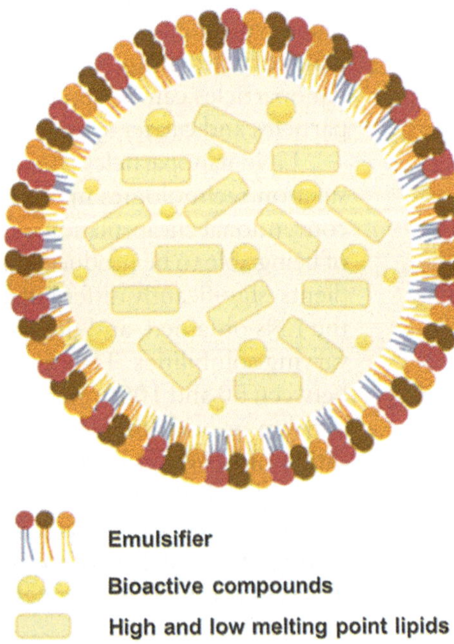

Fig. 1 Schematic representation of the LN structure

as microemulsions, phase inversion temperature, and solvent-based methods (diffusion and solvent injection). These methods are based on the following technological approaches: (i) creation of an oil-in-water nanoemulsion as a precursor to subsequent steps and (ii) subsequent solidification of the dispersed lipid phase. These methods vary substantially in scale but are relatively cost-effective due to the required equipment configuration [2, 3].

2 Materials

2.1 Lipid Nanoparticles Formulation

1. Lipid materials (*see* **Note 1**).
   The selection of lipid materials should consider [1–3]
   a. stability against chemical and enzymatic degradation.
   b. cost-effectiveness and availability.
   c. ability to solubilize, entrap, and protect BC.
   d. biodegradability and non-toxicity.
      i. High melting lipids (*see* **Note 2**).
      ii. Low melting lipids (*see* **Note 3**).
2. Emulsifiers (*see* **Notes 4 and 5**).
3. Co-emulsifiers (*see* **Note 6**).
4. Ultrapure water
5. Lipophilic BC (*see* **Note 7**).

2.2 Lipid Nanoparticles Production

1. High energy approaches:
   a. High-pressure homogenization:
      i. Standard laboratory beakers.
      ii. Temperature-controlled heating plate.
      iii. Ultra-Turrax® high-shear homogenizer.
      iv. High-pressure homogenizer.
   b. Ultrasonic homogenization:
      i. Standard laboratory beakers.
      ii. Temperature-controlled heating plate.
      iii. Ultra-Turrax® high-shear homogenizer.
      iv. Ultrasonic equipment.
      v. Ultrasonic tubes.
      vi. Ultrasonic probe.
      vii. Thermostatic heat bath.

c. Microfluidization:
    i. Standard laboratory beakers.
    ii. Temperature-controlled heating plate.
    iii. Microfluidic devices.

2. Low energy approaches:
    a. Microemulsion:
        i. Standard laboratory beakers.
        ii. Temperature-controlled heating plate.
        iii. Cold water (2–1 °C).
    b. Spontaneous emulsification:
        i. Standard laboratory beakers.
        ii. Temperature-controlled magnetic stirring hot plate.
    c. Solvent diffusion:
        i. Standard laboratory beakers.
        ii. Partially water-miscible organic solvents (e.g., benzyl alcohol, butyl lactate, isobutyric acid, isovaleric acid, and tetrahydrofuran).
        iii. Temperature-controlled magnetic stirring hot plate.
        iv. Ultra-Turrax® high-shear homogenizer.
        v. Ultrafiltration system.
    d. Solvent injection:
        i. Standard laboratory beakers.
        ii. Partially water-miscible organic solvents (acetone, isopropanol, and methanol).
        iii. Temperature-controlled magnetic stirring hot plate.
        iv. Solvent injection needle assembly.
    e. Phase inversion:
        i. Standard laboratory beakers.
        ii. Temperature-controlled magnetic stirring hot plate.
        iii. Cold water.
        iv. Ultra-Turrax® high-shear homogenizer.

2.3 Lipid Nanoparticles Characterization

1. Zeta potential:
    a. Ultrapure water.
    b. Dynamic light scattering (DLS) instrument.
    c. Zeta potential cuvette.

2. Thermal properties:
    a. Differential scanning calorimeter (DSC).
    b. Precision balance (±10 mg accuracy).

   c. Standard aluminum sample pans and lids.
   d. High-purity nitrogen gas.

3. Crystallinity and polymorphism:
   a. X-ray diffractometer with Bragg–Brentano geometry.
   b. Standard aluminum sample capsules.

4. Solid fat content:
   a. Low-field pulsed nuclear magnetic resonance (NMR) spectrometer.
   b. NMR sample tubes.
   c. Temperature-controlled NMR bath.

5. Physical stability:
   a. Turbiscan stability analyzer.
   b. Turbiscan-compatible sample tubes.

3 Methods

3.1 Preparation of Emulsifying Solutions (When Applicable)

1. In beakers, dissolve the emulsifier in ultrapure water under magnetic stirring at room temperature (25 °C).
2. Add sodium azide (0.02%) as an antimicrobial agent, if necessary.
3. Store the solution under refrigeration (5–7 °C) for at least 12 h for complete hydration, if applicable to the type of emulsifier (e.g., proteins).

3.2 Preparation of Lipid Phase

1. Add the desired lipophilic BC (e.g., carotenoids, beta-carotene, tocopherol, essential oil) to the lipid phase.
2. The lipid components must be melted at a temperature of approximately 10°C above the melting point and kept under magnetic stirring until the compound is solubilized for 1 h.

3.3 Preparation of the Pre-Emulsion

1. Heat the lipid phase until complete melting (approximately 85 °C). Lipids with a high melting point (above 30 °C, usually around 80 °C) or a low melting point (below 30°C) can be used (see examples in Sect. 2).
2. Add the emulsifier to the aqueous phase (emulsifiers with high lipophilic hydrophilic balance [HLB]) or to the lipid phase (emulsifiers with low HLB).
3. Heat the aqueous phase to the same temperature as the lipid phase (*see* **Note 8**).

4. Add the aqueous phase to the melted lipid phase under continuous stirring, in the desired ratio (e.g., 90:10 w/w) (*see* **Note 9**).
5. Homogenize the mixture using a high-speed homogenizer (Ultra-Turrax or equivalent) at 5000–10,000 rpm for 3–5 min.

3.4 High-Energy Methods

These techniques use intense mechanical forces to reduce droplet size and promote the formation of stable nanometric particles (*see* **Note 10**).

3.4.1 High-Pressure Homogenization

This procedure was described according to [2, 4] (Fig. 2).

1. Prepare the lipid phase and pre-emulsion according to Sects. 3.2 and 3.3, respectively.
2. Pre-heat the high-pressure homogenizer with hot water (above 70 °C) to prevent lipid crystallization during the homogenization process.
3. Submit the hot pre-emulsion (70–90 °C) to high-pressure homogenization (500–1500 bar for two to five cycles).
4. Cool the nanoemulsions to 5 °C for 24 h to allow crystallization and stabilization of the LN.
5. Store the LN at 25 °C in containers protected from light and oxygen.

3.4.2 Cold-Pressure Homogenization

This procedure was described according to [2, 4] (Fig. 3). In cases where heat-sensitive BC is used, cold homogenization is recommended:

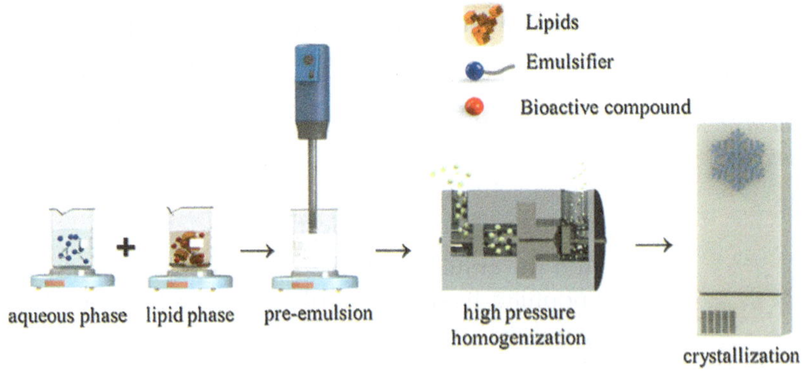

Fig. 2 Hot high-pressure homogenization to produce lipid nanoparticles. (Adapted from [2]. https://doi.org/10.3390/foods14060973, used under a Creative Commons CC BY license)

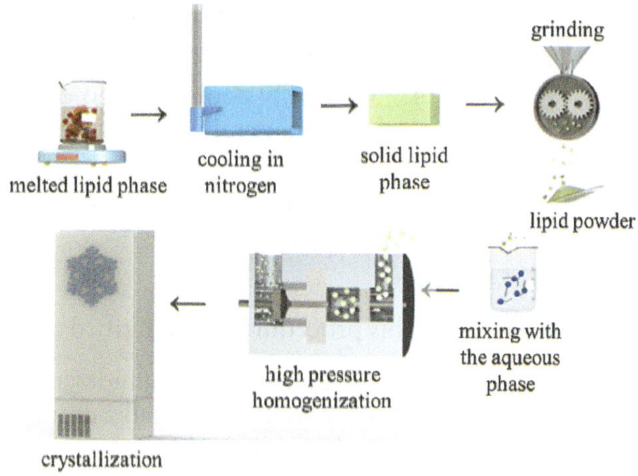

Fig. 3 Cold high-pressure homogenization to produce lipid nanoparticles. (Adapted from [2]. https://doi.org/10.3390/foods14060973, used under a Creative Commons CC BY license)

1. The lipid components must be melted at a temperature of approximately 10 °C above the melting point and kept under magnetic stirring.
2. Rapidly cool the lipid mixture containing the BC using liquid nitrogen or dry ice, forming a solid lipid phase.
3. Grind the frozen lipid mass to obtain micrometric powder particles (approximately 50–100 μm), ensuring the temperature remains below the melting point of the lowest melting lipid.
4. Disperse the ground lipid particles into the aqueous phase (at 4–8 °C) under gentle stirring.
5. Subject the resulting suspension to cold high-pressure homogenization (at room temperature) (500–1500 bar for two to five cycles).
6. Cool the nanoemulsions to 5 °C for 24 h to allow crystallization and stabilization of the LN.
7. Store the LN at 25 °C in containers protected from light and oxygen.

3.4.3 Ultrasonic Homogenization

This procedure was described according to [6–8].

1. Subject the pre-emulsion to high-intensity ultrasound (20–50 kHz) using either direct sonication (ultrasonic probe) or indirect sonication.
2. Adjust the sonication parameters: sonication time (2–8 min), power intensity (≈100 W), cycles (0–50), and amplitude (≈40%) (*see* **Note 11**).

3. After sonication, homogenize again using Ultra-Turrax at 20,000 rpm for 1 minto stabilize the dispersion (optional).
4. Gradually cool the nano emulsion (in an ice bath for 30 min) to a temperature below the crystallization temperature of the lipid.
5. Store the LN at 25°C in containers protected from light and oxygen.

3.4.4 Microfluidization

This procedure was described according to [2, 5].

1. Preheat the microfluidizer with hot water (above 70 °C) to prevent lipid crystallization during the homogenization process (*see* **Note 12**).
2. Feed the pre-emulsion into the high-pressure microfluidizer. For a single-channel device, load the pre-emulsion into the single reservoir (*see* **Note 13**).
3. Immediately cool the emulsion to 4–8 °C after microfluidization to induce lipid phase crystallization and stabilize the LN.
4. Store the dispersion under refrigeration (5 °C) for at least 12 h.

3.5 Low-Energy Methods

3.5.1 Microemulsion

1. Add the lipid phase to the aqueous phase at the same temperature (above the melting point, ≈85 °C). If necessary, the cold microemulsion dilution method can be used (*see* **Note 14**).
2. Shake the system vigorously.
3. For crystallization, disperse the system in cold water (2–10 °C) with a dilution factor between 1:25 and 1:50.
4. If necessary, use ultrafiltration (0.45 μm) or freeze-drying techniques to remove excess water.

3.5.2 Spontaneous Emulsification

Spontaneous emulsification can be achieved by: (i) varying the composition of the phases; (ii) changing the environmental conditions (e.g., temperature, pH, and ionic strength); or (iii) changing the homogenization conditions of the emulsion phases (e.g., order of addition of the components, stirring speed, addition rate) [9–11].

1. Prepare the organic solution by dissolving lipids and the BC along with a lipophilic surfactant in a water-miscible organic solvent (40 mL). Mix using a vortex or magnetic stirrer at 750–1050 rpm for 5 min.
2. Prepare the aqueous phase by adding 80 mL of water containing a hydrophilic surfactant. Stir magnetically at approximately 300 rpm for 15–30 min.

3. In some cases, cooling in an ice-water bath for 2 h is necessary to allow lipid recrystallization.
4. Inject the organic phase into the aqueous phase under magnetic stirring, maintaining agitation for 30 min to allow the system to reach equilibrium [2].
5. Remove the water-miscible solvent by evaporation under reduced pressure for approximately 45 min.
6. Disperse the resulting oil nanodroplets in an aqueous solution containing the hydrophilic surfactant.
7. Allow the dispersions to equilibrate at room temperature.
8. To remove unbound emulsifiers, the dispersions may be transferred into dialysis bags (10–12 kDa molecular weight cutoff, presoaked in ultrapure water for 24 h) and dialyzed against ultrapure water (SLN suspension: water ratio of 1:20 v/v) for 24 h under agitation (300 rpm) at room temperature.
9. The characteristics of the LN—such as PS, PDI, and droplet concentration—are strongly influenced by the relative proportions of the formulation components.

3.5.3 Solvent Diffusion

1. Saturate the partially water-miscible organic solvent with water to ensure initial thermodynamic equilibrium [12].
2. Dissolve the lipids and the BC in the saturated solvent.
3. Emulsify the mixture in an aqueous emulsifier solution saturated with a solvent with a mixer.
4. Gradually add the emulsion to the aqueous phase (emulsion/water ratio of 1:5 or 1:10) under continuous stirring, facilitating the removal of the organic solvent by diffusion and promoting the solidification of the dispersed phase.
5. Transfer to an ice bath (below 5 °C) to solidify the LN.
6. Remove the solvent by ultrafiltration or lyophilization.

3.5.4 Solvent Injection

The type and volume of the injected solvent, the overall concentration of components, the concentration of lipids in the solvent phase, and the viscosity of the aqueous phase can influence and control the particle size of nanostructured lipids [13, 14].

1. Dissolve the lipid phase in a water-miscible solvent (*see* **Note 15**) or in a water-miscible solvent mixture.
2. Prepare the aqueous phase by adding an emulsifier or a mixture of emulsifiers to water or a buffer solution.
3. Inject this mixture slowly into the aqueous phase under constant magnetic stirring (500 rpm) using a syringe with a needle.
4. Remove the solvent by rotary evaporator (45 °C, 80 mbar).

3.5.5 Phase Inversion

1. Mix the ultrapure water and surfactants.
2. Add the lipid phase to the previous mixture, with strict temperature control and gentle magnetic stirring.
3. Increase the system temperature to 10°C above the phase inversion temperature (*see* **Note 16**) under magnetic stirring. The mixture is subjected to cooling and heating cycles to reach the phase inversion zone (e.g., 85–60 °C; 60–85 °C; 85–60 °C). The phase inversion temperature must initially be determined [11, 15].
4. Dilute the system in water at room temperature (25 °C) using three times the volume of water in the hot water-in-oil emulsion. Or dilute in cold water (0 °C) to break the system and promote phase inversion.
5. Store the LN in temperature-controlled incubators at their formation temperatures.
6. When defining formulations, it is necessary to calculate the solvent/oil ratio (Eq. 1) and water/oil weight ratio (Eq. 2). Both are essential for the efficiency of droplet formation and the size of the resulting droplets.

$$\mathrm{SOR} = 100 \times \frac{W_{\mathrm{surfactant}}}{W_{\mathrm{surfactant}} + W_{\mathrm{oil}}} \tag{1}$$

$$\mathrm{WOR} = 100 \times \frac{W_{\mathrm{water}}}{W_{\mathrm{water}} + W_{\mathrm{oil}}} \tag{2}$$

3.6 Lipid Nanoparticles Characterization [1, 16–18]

Particle size (PS), polydispersity index (PDI), entrapment efficiency (EE), and loading capacity (LC) are described in Chapter 7.

3.6.1 Zeta Potential (ZP)

1. Dilutee the samples in ultrapure water (1:100 w/w).
2. Place the samples in appropriate ZP cuvettes.
3. Insert the cuvette in the DLS instrument.
4. The conversion of the electrophoretic mobility to ZP is performed using the Smoluchowski equation (Eq. 5).

$$\zeta = \frac{[\mathrm{EM} \times \eta]}{\varepsilon} \tag{5}$$

where ζ is ZP, EM is electrophoretic mobility, ε is dielectric constant (for water, $\varepsilon = 78.54$), and η is viscosity of the dispersion medium (for water, $\eta = 0.8904$ cp).

3.6.2 Thermal Properties

1. Weigh approximately 10 mg of the sample into 40 μL aluminum pans.
2. Seal the pans hermetically with an appropriate lid.
3. Add the pans to the DSC equipment (modulated DSC 2920).
4. Keep the pans in the equipment in an isothermal condition at 25 °C for 10 min for thermal stabilization.
5. After equilibration, subject the samples to a heating cycle from 25 to 100 °C, with a heating rate of 10 °C/min.
6. Then cool to 25 °C.
7. Determine the analysis parameters (initial melting temperature, peak melting temperature, enthalpy of melting, and final melting temperature) using Universal Analysis 2000 software (TA Instruments).

3.6.3 Crystallinity and Polymorphism

1. Prepare the dry sample, ensuring a uniform and flat surface for analysis.
2. Position the sample in the sample holder of the X-ray diffractometer [18].
3. Set up the equipment with the following parameters: acceleration voltage: 40 kV, tube current: 30 mA, scan range (2θ): 5°–40°, step size: 0.02°, acquisition time per step: 2 s, measurement geometry: Bragg–Brentano (θ–2θ) with Cu-Kα radiation ($\lambda = 1.54056$ Å).
4. Perform the scan in the range of 5°–40° (2θ).
5. Perform the measurements with steps of 0.02° in 2θ and an acquisition time of 2 s, with scans from 5° to 40° (scale 2θ).
6. Identify the polymorphic forms based on the characteristic short-spacing peaks observed in the diffractogram.
7. Estimate the relative proportion between the different crystalline forms from the relative intensity of these peaks.
8. Use specific scientific literature to confirm the detected polymorphic phases.

3.6.4 Solid Fat Content

1. Transfer the samples to NMR tubes after recrystallization of the LN [18].
2. Place the tubes in the Bruker pc120 Minispec low-resolution pulsed NMR spectrometer, using a high-precision dry bath at 25 °C.
3. Readings will be recorded every 1 min until complete stabilization.

3.6.5 Physical Stability

1. Place the samples in Turbiscan compatible tubes at 25 °C
2. Detection is performed in a Turbiscan optical analyzer using an infrared light source with a wavelength of 880 nm, with transmission and backscatter detectors.
3. Using the TurbiSoft 2.0 software, the emulsion stability index is determined, and possible destabilization mechanisms are identified: cremation, sedimentation, flocculation, or coalescence.

4 Notes

1. There is a growing use of edible or commercially available oils and fats instead of synthetic lipids, prompted by economic and regulatory considerations. Nanoparticle technology in the food industry now emphasizes biocompatible lipid fractions sourced from natural oils and fats, along with emulsifiers that are renewable and approved for food use. Lipid systems are selected according to their chemical composition and crystallization properties, which support nanostructure development. The focus on conventional lipid materials from the oil and fat sector offers potential for broader application of nanoparticles in food products.
2. High-melting-point purified lipids (e.g., stearic acid, tristearin, tripalmitin, trilaurin, trimyristin, glyceryl behenate, glyceryl stearate); fully hydrogenated vegetable oils (e.g., soybean, sunflower, rapeseed, palm); natural waxes (e.g., candelilla, carnauba, bees, rice bran, sunflower); natural fats (e.g., cocoa butter, coconut fat, milk fat, murumuru butter).
3. Low-melting-purified lipids (e.g., oleic acid, medium-chain triacylglycerols); vegetable oils (e.g., soybean, sunflower, corn, rapeseed, high-oleic sunflower, and soybean oils); fish and microalgae oils.
4. Examples:
   - Ionic emulsifiers (e.g., sodium lauryl sulfate, citric acid esters of monoacylglycerols, diacetyl tartaric esters of fatty acids and glycerol, lauric alginate).
   - Non-ionic emulsifiers (e.g., Tween 20, Tween 60, Tween 80, Span 80, Span 60, poloxamer 188, mono- and diacylglycerols of fatty acids).
   - Phospholipids such as lecithins (e.g., soybean, egg, and sunflower lecithins), phosphatidylcholine, lysophospholipids, glycolipids (e.g., rhamnolipids).
   - Protein emulsifiers: plant-based protein isolates (e.g., soybean, pea); animal-based protein isolates (e.g., casein, whey).

5. The choice of emulsifier for lipid nanoparticles depends on the ability to form stable systems, HLB value, processing conditions, and final application. Very weak surfactants may not adequately reduce interfacial tension, while excessively strong surfactants may destabilize the emulsion over time. Their properties can be adjusted by varying the type and concentration, but they remain sensitive to factors such as temperature, pH, and the presence of ions. The first step in selection is to determine whether a natural or synthetic emulsifier will be used. Ionic emulsifiers and proteins stabilize through electrostatic repulsion, thus being more sensitive to variations in pH and ionic strength. However, proteins are considered natural emulsifiers. Non-ionic emulsifiers are the most versatile, providing steric stabilization. Phospholipids, such as lecithin, efficiently reduce interfacial tension through both steric and electrostatic stabilization [19].
6. Sodium dodecyl sulfate, sodium taurodeoxycholate, sodium glycolate.
7. Carotenoids (e.g., beta-carotene, astaxanthin, lutein, zeaxanthin); phytosterols; tocopherols; curcumin; fat-soluble vitamins (A, D, E, and K); omega-3 fatty acids (e.g., EPA and DHA); essential oils (e.g., thyme, cinnamon); cannabidiol.
8. In cases of hydrophilic BC, insert them at this stage.
9. To achieve a 90:10 (w/w) ratio in the preparation of pre-emulsions, simply remember that this proportion represents the mass fraction that each phase should occupy in the system. Thus, 90% of the total mass corresponds to the lipid phase (including molten lipids and any surfactant or lipophilic compound dissolved in it) and 10% corresponds to the aqueous phase (water, water-soluble solutes, and hydrophilic surfactants). To calculate, first define the desired total mass and multiply it by 0.90 to obtain the amount of lipid phase and by 0.10 to obtain the amount of aqueous phase. Since the ratio is by mass, not volume, the components must be weighed or volumes converted using density.
10. The choice of LN production method depends on the application and whether a high- or low-energy process is required. High-pressure homogenization is popular for its scalability and equipment availability, while microfluidization offers uniformity, reproducibility, low thermal impact, minimal nutrient loss, and fast production—making it ideal for heat-sensitive formulations. Both methods have distinct advantages and should be selected based on specific needs, compound properties, scalability, and cost.
11. Care should be taken to avoid overprocessing, which can cause instability and/or thermal degradation of sensitive

compounds. Sonication promotes PS reduction through cavitation and shearing, forming the LN.

12. During processing, the emulsion is divided into two streams that collide in an interaction chamber, promoting a reduction in droplet size due to shear forces, cavitation, and pressure drop. The following can be used: single channel (indicated for more homogeneous systems) or double channel (for complex emulsions or those with a high lipid-phase content). The single channel has a single inlet reservoir, requiring the prior formation of the pre-emulsion. The double channel has two inlet reservoirs, allowing simultaneous feeding of the aqueous and lipid phases, thus eliminating the pre-emulsion stage.
13. This step should be used when working with high-melting-point lipids. For a dual-channel device, load the lipid phase and aqueous phase separately into their respective reservoirs.
14. Perform this step carefully as it may modify the physical characteristics of the LN.
15. Examples of miscible solvents in water are: benzyl alcohol, butyl lactate, isobutyric acid, isovaleric acid, tetrahydrofuran, methanol, ethanol, isopropanol, or acetone.
16. Temperature changes modify the physicochemical properties (solubility and molecular geometry) of non-ionic surfactant agents, promoting phase inversion. The main limitation of this method is that only BC resistant to high temperatures can be incorporated. This temperature can be determined by monitoring the change in turbidity in a UV/visible spectrophotometer at 600 nm of the surfactant–oil–water systems during heating at a constant rate [15].

References

1. Tamjidi F, Shahedi M, Varshosaz J, Nasirpour A (2013) Nanostructured lipid carriers (NLC): a potential delivery system for bioactive food molecules. Innov Food Sci Emerg Technol 19:29–43. https://doi.org/10.1016/j.ifset.2013.03.002
2. Lüdtke FL, Silva TJ, da Silva MG et al (2025) Lipid nanoparticles: formulation, production methods and characterization protocols. Foods 14:937. https://doi.org/10.3390/foods14060973
3. da Silva Santos V, Badan Ribeiro AP, Andrade Santana MH (2019) Solid lipid nanoparticles as carriers for lipophilic compounds for applications in foods. Food Res Int 122:610–626. https://doi.org/10.1016/j.foodres.2019.01.032
4. Yang Y, Corona A, Schubert B et al (2014) The effect of oil type on the aggregation stability of nanostructured lipid carriers. J Colloid Interface Sci 418:261–272. https://doi.org/10.1016/j.jcis.2013.12.024
5. Kavinila S, Nimbkar S, Moses JA, Anandharamakrishnan C (2023) Emerging applications of microfluidization in the food industry. J Agric Food Res 12:100537. https://doi.org/10.1016/j.jafr.2023.100537
6. Osanlou R, Emtyazjoo M, Banaei A et al (2022) Preparation of solid lipid nanoparticles and nanostructured lipid carriers containing zeaxanthin and evaluation of physicochemical properties. Colloid Surf A Physicochem Eng Asp 641:128588. https://doi.org/10.1016/j.colsurfa.2022.128588
7. Behbahani ES, Ghaedi M, Abbaspour M, Rostamizadeh K (2017) Optimization and characterization of ultrasound assisted preparation

of curcumin-loaded solid lipid nanoparticles: application of central composite design, thermal analysis and X-ray diffraction techniques. Ultrason Sonochem 38:271–280. https://doi.org/10.1016/j.ultsonch.2017.03.013

8. Pucek-Kaczmarek A (2021) Influence of process design on the preparation of solid lipid nanoparticles by an ultrasonic-nanoemulsification method. Processes 9:1265. https://doi.org/10.3390/pr9081265
9. Bouchemal K, Briançon S, Perrier E, Fessi H (2004) Nano-emulsion formulation using spontaneous emulsification: solvent, oil and surfactant optimisation. Int J Pharm 280:241–251. https://doi.org/10.1016/j.ijpharm.2004.05.016
10. Shirvani A, Goli SAH, Varshosaz J et al (2022) Fabrication of edible solid lipid nanoparticle from beeswax/propolis wax by spontaneous emulsification: optimization, characterization and stability. Food Chem 387:132934. https://doi.org/10.1016/j.foodchem.2022.132934
11. Anton N, Vandamme TF (2009) The universality of low-energy nano-emulsification. Int J Pharm 377:142–147. https://doi.org/10.1016/j.ijpharm.2009.05.014
12. Mousavi-Simakani SM, Azadi A, Tanideh N et al (2023) Simvastatin-loaded nanostructured lipid carriers as topical drug delivery system for wound healing purposes: preparation, characterization, and *in vivo* histopathological studies. Adv Pharm Bull 13:761–771. https://doi.org/10.34172/apb.2023.083
13. Duong VA, Nguyen TTL, Maeng HJ (2020) Preparation of solid lipid nanoparticles and nanostructured lipid carriers for drug delivery and the effects of preparation parameters of solvent injection method. Molecules 25:1–36. https://doi.org/10.3390/molecules25204781
14. Uchôa AFC, Formiga ALD, Alves ÁEF, et al (2025) Optimization and functionalization of copaiba oil-loaded nanostructured lipid carriers to improve cytotoxicity against breast cancer cells. J Drug Deliv Sci Technol 105, 106575. https://doi.org/10.1016/j.jddst.2024.106575
15. Gao S, McClements DJ (2016) Formation and stability of solid lipid nanoparticles fabricated using phase inversion temperature method. Colloid Surf A Physicochem Eng Asp 499:79–87. https://doi.org/10.1016/j.colsurfa.2016.03.065
16. Stahl MA, Lüdtke FL, Grimaldi R et al (2024) Characterization and stability of solid lipid nanoparticles produced from different fully hydrogenated oils. Food Res Int 176. https://doi.org/10.1016/j.foodres.2023.113821
17. Shah RM, Malherbe F, Eldridge D et al (2014) Physicochemical characterization of solid lipid nanoparticles (SLNs) prepared by a novel microemulsion technique. J Colloid Interface Sci 428:286–294. https://doi.org/10.1016/j.jcis.2014.04.057
18. American Oil Chemists' Society (AOCS) (2009) Official methods and recommended practices of the American Oil Chemists' Society, 6th edn, AOCS Press, Champaign, IL.
19. Henao-Ardila A, Quintanilla-Carvajal MX, Moreno FL (2024) Emulsification and stabilisation technologies used for the inclusion of lipophilic functional ingredients in food systems. Heliyon 10(11):e32150. https://doi.org/10.1016/j.heliyon.2024.e32150

Chapter 8

Internal and External Ionic Gelation

Maria Clara Guimarães and Ana Silvia Prata

Abstract

Ionic gelation (IG) is a versatile encapsulation method that can be applied to both hydrophilic and lipophilic compounds. It utilizes the interaction between polyelectrolytes and oppositely charged ions to form three-dimensional gels. With broad applications in food, pharmaceuticals, and cosmetics, IG allows the encapsulation of chemical and biological agents, bioactive compounds, and flavorings for controlled release. Its advantages include ease of execution, low cost, reproducibility, and large-scale industrial applicability. This protocol details the preparation of beads formation with hydrophobic core materials by dripping while also addressing adaptations for simple and double emulsions, as well as direct and reverse gelation processes. This technique is promising because of its versatility and ability to protect and release compounds in a controlled manner across various applications.

Keywords Ionotropic gelation, Extrusion, Internal gelation, External gelation, Essential oil

1 Introduction

Ionic gelation, also known as ionotropic or ion-induced gelation, uses mild and aqueous processing conditions, non-toxic reagents, and requires very low energy which make this production route the best suited for many biological application but also for other correlate areas. [1]. The technique is based on polyelectrolytes (chitosan, alginate, hyaluronic acid, and carrageenan) capable to being cross-linked in the presence of counterions, and as consequence to transition of sol-gel. In ionic cross-linking, the formation of a network occurs through ionic bridges between macromolecular chains. The cross-linking agent is usually a charged ionic entity with a well-defined molecular weight [2]. Spherical beads or the smallest particles containing a core material can be obtained by dripping or spraying a polymeric dispersion onto a charged crosslinker solution, through a syringe or high-pressure nozzle, respectively (Fig. 1).

The gelation can be performed in two ways: (1) External gelation: The polymer solution is added dropwise to a solution containing ions. The gel network is formed as ions diffuse from outside (ion

Ana Silvia Prata and Louise Emy Kurozawa (eds.), *Biopolymers*, Methods and Protocols in Food Science,
https://doi.org/10.1007/978-1-0716-5230-5_8,

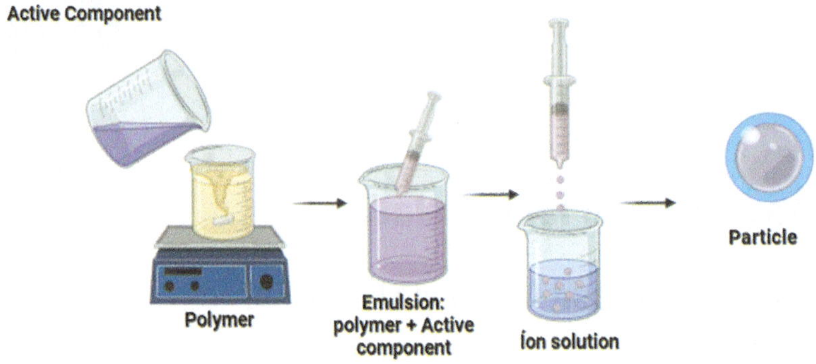

Fig. 1 Ionic gelation technique. (Figure created by the author using Biorender)

solution) to inside (polymer solution), and then thee external layer is immediateley formed. The the sol–gel transition process is fast or (2) Internal gelation: Ions are added into a solution/emulsion, and then dropped into the hydrocolloid system to form the hydrogel beads or microparticles. Gelation occurs from the inside out. Since typically used with emulsions the resulting soft molecular shell features depend on the type of emulsion used (W/O or O/W).

By using solution or different emulsions - simple, double, W/O, and O/W the IG can encapsulate both types of core materials, hydrophillic and lipophilic ones [4]. IG is particularly significant for encapsulation and controlled release of sensitive core materials. Despite hydrosoluble core material including probiotics, biotin, insulin, ascorbic acid, antoccyanins, cafein [3–6] to be entrapped by hydrogel matrices, many challenges linked to the high affinity of core material to the aqueous media must be overcome. On the. other hand, lipophilic compounds are incorporated into the hydrogels with higher retention [7–9].

The main disadvantages of the ionotropic bead are the low encapsulation efficiency and rapid release of the highly water-soluble core materials due to the large pore sizes formed, which make trapping small molecules difficult. One solution to this problem is the combination of polymers with other compounds, which can help reduce pore size and improve the encapsulation efficiency of smaller molecules [10]. In addition, mechanical instability can be a problem because, depending on the polymer used, the capsules can be fragile and break under physical stress.

2 Materials

1. Hydrophilic core material: essential oil (*see* **Note 1**).
2. Calcium chloride ($CaCl_2$) (*see* **Note 2**).

3. Sodium alginate (*see* **Note 3**).
4. Tween 80.
5. Syringe with a fine needle, 6 mm × 0.25 mm (*see* **Note 4**).
6. Beaker or container for preparing the solutions.
7. Distilled water.
8. Magnetic stirrer with temperature control.
9. Deionized water.
10. pH meter.

3 Methods

3.1 External Gelation

1. Dissolve the polymer in distilled or purified water at 1% w/v, under continuous stirring in a hot plate agitator at 50°C.
2. The solution is allowed to stand for 1 h at ambient temperature to remove air bubbles.
3. Prepare a gelling solution by dissolving the calcium chloride at 0.5 M in distilled water.
4. Place the gelling solution in a 300 mL beaker under gentle magnetic stirring.
5. Prepare the experiment setup by placing the syringe in a vertical position over the glass beaker containing the gelling solution (Fig. 1) (*see* **Note 5**).
6. Incorporate the core material into the polymer solution at ambient temperature and homogenize to obtain an emulsion (eg. 14,0000 rpm for 3 min). When working with hydrophilic core material, just mix it with the polymer solution to achieve a homogeneous solution or suspension.
7. Transfer the polymeric phase to a burette or syringe.
8. Slowly add dropwise this solution to the gelling ion solution, while maintaining a constant magnetic stirring.
9. Beads should stay at least 15 min in the $CaCl_2$ solution
10. Sieve the beads with a mesh smaller than their sizes and wash them with water.
11. Drying is optional (*see* **Note 6**).

3.2 Internal Gelation

1. Prepare an emulsion with the 5 mL of essential oil, 1.5 mL of $CaCl_2$ (40 g/L), and 0.025 mL of Tween 80, homogenizing for 1 min with Ultra-Turrax at 14,000 rpm.
2. Prepare 150 mL of the gelling solution by dissolving the alginate in distilled water at 1% w/v, under continuous stirring in a hot plate agitator at 50°C.

3. Place this solution in a 300 mL beaker and keep rotating at 200 rpm on a magnetic stirrer at room temperature
4. Prepare the experiment setup by placing the syringe in a vertical position over the glass beaker containing the gelling solution (Fig. 1).
5. Transfer the emulsion to a burette or syringe.
6. Slowly add dropwise the emulsion to the gelling ion solution, while maintaining a gentle and constant stirring.
7. Sieve the beads with a mesh smaller than their sizes and wash them several times with deionized water containing 0.1% v/v Tween 80.
8. Drying is optional (*see* **Note 6**).

The particle should be evaluated for size and encapsulation efficiency as described in Chap. 17.

4 Notes

1. If using a hydrophobic core material, it is required to emulsify it in an O/W emulsion. An O/W primary emulsion may be prepared by blending hydrophobic compounds (up to 2%) with the alginate solution (containing 1% Tween 80 (w/v) or not). W/O/W-multiple emulsion may be prepared first by dispersing the aqueous phase in oil (4:6 v/v) using a surfactant with very low HLB (eg. PGPR, 3% w/v) which results in an W/O emulsion.This emulsion is reemulsified with alginate solution (1% w/v) in a lower shear rate. In both cases, alginate is the external phase. and prone to be useed with direct gelation.
2. Calcium ions (Ca^{2+}) are frequently employed with alginates and pectins. Other divalent and trivalent ions can also be employed, and the size of these ions influences the resistance of the gel network. Tripolyphosphate (TPP) is a common anion used with chitosan.
3. Mannuronic and Guluronic groups balance influences the gelling properties and the viscosity of the polymeric solution. A complete study can be seen in [11]. For pectins, the low methoxylates are more indicated with maintaining an appropriate pH. For chitosan, attention should be paid to the deacetylation degree.
4. Beyond dripping by gravity, the size of beads can be reduced by using peristaltic pumping (up to 1 mm), Jet Cutter (up to 600 um), or pressurized air (smaller than 200 um).

5. The height between the tip and the bath influences the format of the bead. Fix the dripping distance for example at 10-30 cm of. the tip above the surface of the gelling bath.
6. The beads can be dried in a convection oven to reduce their humidity content. Alternatively, the beads can be frozen at−80°C for 2 h and lyophilized for 14 h.

References

1. Sreekumar S, Goycoolea FM, Moerschbacher BM et al (2018) Parameters influencing the size of chitosan-TPP nano- and microparticles. Sci Rep 8:4695. https://doi.org/10.1038/s41598-018-23064-4
2. Fernández-Quiroz D, Tohidi MM, Paymard B, Lucero-Acuña A (2023) Chapter 9: Immobilization of essential oils in biopolymeric matrices: recent approaches for controlled delivery systems. In: Rahman A (ed) Studies in Natural Products Chemistry: Vol 78. Elsevier pp 365-401. ISBN 9780323912532. https://doi.org/10.1016/B978-0-323-91253-2.00005-4
3. Silveira MP, Francisco LCA, Alvim ID, Prata AS (2023) Encapsulation of pomegranate polyphenols by ionic gelation: strategies for improved retention and controlled release. Food Res Int 174:113590. https://doi.org/10.1016/j.foodres.2023.113590
4. Yun P, Devahastin S, Chiewchan N (2021) Microstructures of encapsulates and their relations with encapsulation efficiency and controlled release of bioactive constituents: a review. Compr Rev Food Sci Food Saf 20(2):1768–1799. https://doi.org/10.1111/1541-4337.12701
5. Mosquera-Vivas E, Ayala-Aponte A, Serna-Cock L, Torres-León C, Tirado DF (2024) Viability of Lactobacillus reuteri DSM 17938 encapsulated by ionic gelation during refractance window® drying of a strawberry snack. Foods 13(6):823. https://doi.org/10.3390/foods13060823
6. Razavi S, Janfaza S, Tasnim N, Gibson DL, Hoorfar M (2021) Microencapsulating polymers for probiotics delivery systems: preparation, characterization, and applications. Food Hydrocolloids 120:106882. https://doi.org/10.1016/j.foodhyd.2021.106882
7. Fujiwara, GM, Campos R, Costa C, Dias J, Obdulio M, Marilis M, Marques F, Zanin S (2013) Production and characterization of alginate-starch-chitosan microparticles containing stigmasterol through the external ionic gelation technique. Brazil J Pharm Sci 49(3):537–547. https://doi.org/10.1590/S1984-82502013000300015
8. Geng M, Li L, Tan X, Teng F, Li Y (2024) W/O/W emulsion-filled sodium alginate hydrogel beads for co-encapsulation of vitamins C and E: Insights into the fabrication, lipolysis, and digestion behavior. Food Chem 457:140095. ISSN 0308–8146. https://doi.org/10.1016/j.foodchem.2024.140095
9. Benavides S, Cortés P, Parada J, Franco W (2016) Development of alginate microspheres containing thyme essential oil using ionic gelation. Food Chem 204:77–83. ISSN 0308-8146. https://doi.org/10.1016/j.foodchem.2016.02.104
10. Kahraman S, Demircan H, Sarioglu K, Demircan S, Oral RA (2023) Investigation of the effects of some wall materials and encapsulation techniques on the microencapsulation of olive leaf extract. Acta Alimentaria 52(3):309–400. https://doi.org/10.1556/066.2022.00253
11. Bhujbal SV, Paredes-Juarez GA, Niclou SP, de Vos P (2014) Factors influencing the mechanical stability of alginate beads applicable for immunoisolation of mammalian cells. J Mech Behav Biomed Mater 37:196–208. https://doi.org/10.1016/j.jmbbm.2014.05.020

Chapter 9

Microencapsulation of Lipophilic Compounds by Complex Coacervation

Jorge A.D. Duarte, Nathalia D. Gonçalves and Ana Silvia Prata

Abstract

Complex coacervation is a microencapsulation technique in which two oppositely charged polyelectrolytes, under diluted conditions, form a phase-separated polymeric material that can be able to encapsulate oils or hydrophobic substances. This chapter aims to describe the steps involved in the complex coacervation process to create oil-containing particles. The method can be optimized by adjusting polymer concentrations, emulsification conditions, and salt concentrations, providing flexibility in the size and normally high encapsulation efficiency. These particles can be further used in controlled release applications, food processing, or drug delivery.

Keywords Coacervates, Active ingredient, Electrostatic interaction, Biopolymers, Gelatin, Gum Arabic

1 Introduction

Complex coacervation is a process whereby two biopolymers of opposite charges, typically polysaccharides and proteins, interact electrostatically in suspension, forming an immiscible liquid phase (coacervates) [1–3].

Encapsulation occurs when, by emulsifying a core compound in a polymeric phase and inducing the complexation with the second polymer, a coacervate phase is formed around the active ingredient, resulting in the sedimentation of the encapsulated core. This method diverges from simple coacervation that occurs by dehydration of a solution containing a single polymer with the addition of a salt or liquid solvent (e.g., Na2SO4, ethanol) [4].

Since particle formation depends on the charge density, attention should be paid to parameters such as polymer concentration, type of polymers, ionic strength, and pH of the solution [3, 4, 5]. Then, a preliminary step is required to understand the complexation conditions of a polymeric pair [3]. In general terms, it is necessary

Ana Silvia Prata and Louise Emy Kurozawa (eds.), *Biopolymers*, Methods and Protocols in Food Science,
https://doi.org/10.1007/978-1-0716-5230-5_9,

to establish the ideal conditions to achieve charge neutrality of both polyelectrolytes within the system.

Usually, complex coacervation is favorable for hydrophobic compounds, and known to provide high encapsulation efficiencies, to be a simple technique in terms of equipment requirements, and useful for volatile and heat-sensitive bioactive compounds, because it does not require high temperatures or harsh solvents [3, 6].

The resulting microparticles are insoluble in water present around 80%–90% of water content, and their diameters vary between 10–80 μm. Nevertheless, these particles are susceptible to alterations in pH and ionic strength [4], and further enzymatic, ionic, or chemical crosslinking step may be necessary to stabilize the complex.

This chapter provides a step-by-step of the complex coacervation microencapsulation of lipophilic compounds using the classic polymeric pairs gum Arabic—gelatin.

2 Materials

1. Cationic Polyelectrolyte: bovine skin gelatin type B (*see* **Note 1**).
2. Anionic Polyelectrolyte: Gum Arabic (*see* **Note 2**).
3. Oil phase: Sunflower oil with paprika resin (10:1) (*see* **Notes 3** and **4**).
4. Deionized water.
5. Hydrochloric acid P.A. (HCl) at 1 and 0.1 M.
6. Sodium hydroxide at 1 and 0.1 M (if necessary).
7. Magnetic stirrer with heat control or overhead stirrer.
8. Cylindrical magnetic bar (7 × 25 mm).
9. High-speed homogenizer or sonicator (for emulsification).
10. Stainless steel digital spike thermometers.
11. Benchtop pH meter with electrode and temperature sensor. The pH meter must be previously calibrated.
12. Analytical balance (precision 0.001 g).
13. Glass beakers and flasks (250 mL, 600 mL, and 1000 mL).
14. Zetasizer device.

3 Methods

3.1 Preparation of Polymeric Solutions

1. Dissolve, separately, both polymers in deionized water to a concentration of 2.5% (w/w) (*see* **Note 5**).

2. Stir the solution in a magnetic stirrer for at least 4 h to ensure complete dissolution. Gelatin solution requires heating to 50°C.

3.2 Defining the Interaction Conditions by Surface Charge Density

1. Dilute the polymeric solutions into ultrapure water to obtain 650 mL of a solution of 0.1% (w/v).
2. Distribute the solution into beakers of 50 mL, totaling 16 beakers containing 40 mL each one.
3. Stir the solutions in a magnetic stirring with enough speed to generate a vortex effect (200 to 300 rpm).
4. Assemble a 5 mL burette in a universal support for dripping aliquots (0.5 or 1 mL) of acid or base solutions (0.1 or 1 M) in the beakers (50 mL).
5. Adjust the pH of the solutions, in duplicates, to the pHs 3.5, 4.0, 4.5, 5.0, 6.0, 7.0, 8.0, and 9.0.
6. Transfer, with a 1 mL pipette, each solution in the different pH values into a disposable cuvette with a capillary cell to measure the zeta potential (*see* **Notes 6, 7 and 8**).
7. Based on the data collected, plot a curve of the average ζ potential data from five measurements (Y-axis) as a function of pH values (X-axis). From this graph, determine the ratio between the polymers at a given pH to achieve charge neutrality (e.g., for gelatin type B and gum Arabic, at pH 4.0, the ratio is 1).

3.3 Complex Coacervation

1. Weigh the oil (2.500 g) in a 250 mL plastic beaker. Before starting the coacervation process, it is necessary to maintain the temperature of the polymeric solutions at 40°C in the stirring plate with constant shaking (*see* **Note 9**; Fig. 1).
2. Combine the oil phase with 100 mL of the first polymeric phase (Gelatin solution, 2.5% (w/w)).
3. Emulsify the oil into fine droplets by using a high-speed homogenizer (typically 14,000 rpm for 3 min) or a sonicator.
4. In a previously weighted 1000 mL beaker, add 100 mL of the anionic polymer solution (gum Arabic solution, 2.5% (w/w)), and then, gradually, add the oil-water emulsion (*see* **Note 10**).
5. Keep the system under constant stirring with enough speed to generate a vortex effect.
6. Adjust the pH of the final suspension to pH 4.0 with HCl solution (0.1 or 1 M) to induce coacervation.
7. Keep under stirring to allow complete droplet separation and insert 600 mL of deionized water, also at 40°C.
8. Re-check the pH and adjust it to pH 4.0, if necessary.

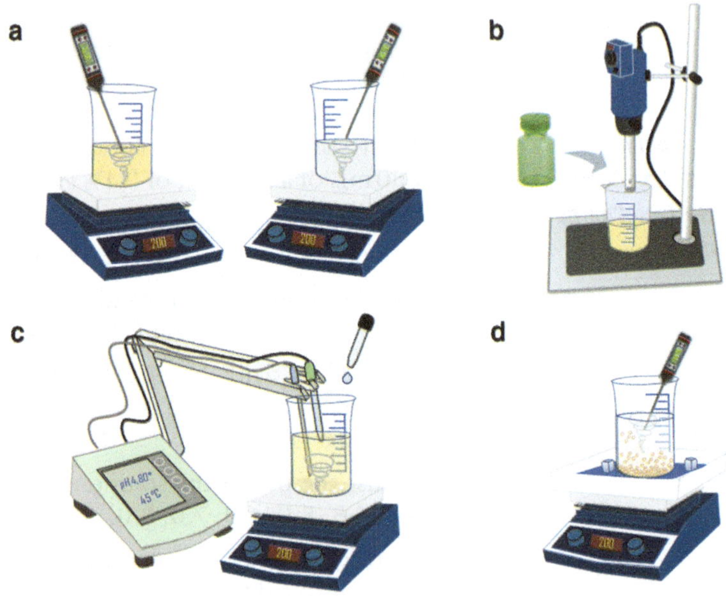

Fig. 1 Schema of the complex coacervation process. (**a**) Prepare both polymeric phases with proper temperature, (**b**) homogenization of the emulsifying phase (protein) with the oily phase; (**c**) pH adjustment of the polymer dispersion solution to induce coacervation; (**d**) complexation and settling of the particles in a cold bath (Source: Author's work)

9. Prepare an ice bath in a polypropylene container measuring 225 × 180 × 4 mm
10. Put it over the stirring and insert the beaker containing the coacervate system to cool it to 10°C.
11. Keep the system under stirring, ensuring that the coacervate phase forms fully.
12. After cooling, the beaker containing the microparticles is kept in a refrigerator for 16 h for phase separation.

3.4 Evaluation of the Efficacy of the Coacervation Process

1. Separate the coacervates settled at the bottom by meticulous pouring off of the water phase (supernatant), to avoid the loss of colloidal material. Wait some hours for additional compaction of the colloidal phase and remove the final few milliliters of the aqueous phase with a pipette (*see* **Note 11**).
2. Weigh the beaker containing the moist particles to determine the amount of coacervates formed. A water film should be kept over particles.
3. Transfer the coacervates to a closed recipient and keep them in the refrigerator for further use.

4. Determine the moisture content of the coacervates in triplicate (*see* **Note 12**).
5. Calculate the yield of the coacervation process by using (Eq. 1).

$$Coacervation\ yield = \frac{(Coacervate\ mass * \%\ TS)}{(Initial\ solids\ mass)}\ x\ 100 \qquad (1)$$

where initial mass of solids corresponds to the oil (2.500 g) and polymers (5.000 g) initially added to the system. %TS is the total solid content calculated from the moisture determination of the coacervate phase. Another important characteristic of these particles is the load—the ratio between the mass of encapsulated oil per gram of wet/dry particle. Moisture, load, encapsulation efficiency, and size determination are described in Chap. 17.

The efficiency of the complexation process (EPC) can be calculated as

$$EPC = \frac{(Coacervate\ mass - (Coacervate\ mass\ *\ Load))}{(Initial\ solids\ mass)}$$

4 Notes

1. All proteins show positive charges below their Isoelectric point (IP). Examples of polymers that can be employed for cationic contribution: protein from plant sources (e.g., pea, wheat, lentil, and soybean), animal-based sources (e.g., whey, egg albumin, silk fibroin, and gelatin), and chitosan, a known cationic polysaccharide.
2. Most polysaccharides are anionic under their pKa, and therefore contribute negatively with charges to the system over a large range of pH. Examples: agar, carboxymethylcellulose, pectin, carrageenan, modified starch, and alginate.
3. Any lipophilic compound, such as essential oils, plant oils (e.g., olive oil, sunflower oil), or other hydrophobic compounds (e.g., paraffin oil, vitamin B12, vitamin D3).
4. The characteristic of the oil changes the ability of the coacervate to encapsulate it. If it lacks emulsification powder in the polymers, the use of surfactant with low molecular weight may help to emulsify the oil phase (e.g., Span 80, Tween 80). If surfactant is needed, normally concentrations of 0.1%–1% (w/v) in the oil phase are adopted.
5. Take a note of the exact weight of polymers and water.

6. Before inserting the cuvette into the zetasizer tray, make sure that no bubbles have been produced inside the capillary path. Also, make sure that the cuvette electrodes are not wet with the solutions.
7. If the same cuvette is used for zeta potential measurements, wash it with deionized water circulating in the capillary at least 3 times.
8. The surface charge density of the polymer is determined by measuring the zeta potential. The electrophoretic light scattering (ELS) technique used in the Zetasizer calculates the electrophoretic mobility of the scattering particles and converts this into zeta potential (mV).
9. The amount of core material can vary from 5% to 150% (w/w) in relation to the wall material (dry basis). For 5 g of polymers, 2.5 g of oil represents 50% of the oil. The temperature used in the system depends on the type of polymer involved.
10. The addition should be done slowly to avoid disturbing the stabilization of oil droplets. The ratio used in this case was 1:1 (100 ml of both solutions at the same concentration).
11. Some systems should be centrifuged at 3000–5000 rpm for 10–15 min to separate the coacervate phase.
12. About 3 g of the wet microparticles is normally used for the determination of moisture content. For essential oils or volatile compounds adopt a vacuum oven at 50°C for 12 h [7–13].

References

1. Thies C (2016) Encapsulation by complex coacervation. In: Encapsulation and controlled release technologies in food systems. Wiley, pp 41–77
2. Hernández-Nava R, López-Malo A, Palou E et al (2020) Encapsulation of oregano essential oil (*Origanum vulgare*) by complex coacervation between gelatin and chia mucilage and its properties after spray drying. Food Hydrocoll 109:106077
3. Prata AS, Grosso CRF (2015) Influence of the oil phase on the microencapsulation by complex coacervation. 92:1063 - 1072
4. Timilsena YP, Akanbi TO, Khalid N et al (2019) Complex coacervation: principles, mechanisms and applications in microencapsulation. Int J Biol Macromol 121:1276–1286
5. Schmitt C, Turgeon SL (2011) Protein/polysaccharide complexes and coacervates in food systems. Adv Colloid Interface Sci 167:63–70
6. Dima C, Cotârlet M, Alexe P et al (2014) Microencapsulation of essential oil of pimento [*Pimenta dioica* (L) Merr.] by chitosan/k-carrageenan complex coacervation method. 22:203–211
7. Devi N, Kakati DK (2013) Smart porous microparticles based on gelatin/sodium alginate polyelectrolyte complex. J Food Eng 117:193–204
8. Jun-xia X, Hai-yan Y, Jian Y (2011) Microencapsulation of sweet orange oil by complex coacervation with soybean protein isolate/gum Arabic. Food Chem 125:1267–1272
9. Girardi NS, García D, Passone MA et al (2017) Microencapsulation of *Lippia turbinata* essential oil and its impact on peanut seed quality preservation. Int Biodeterior Biodegradation 116:227–233
10. Gonçalves ND, de L Pena F, Sartoratto A et al (2017) Encapsulated thyme (Thymus vulgaris)

essential oil used as a natural preservative in bakery product. 96:154–160
11. Santos MB, Geraldo de Carvalho M, Garcia-Rojas EE (2021) Carboxymethyl tara gum-lactoferrin complex coacervates as carriers for vitamin D3: encapsulation and controlled release. Food Hydrocoll 112:106347
12. Xiao Z, Liu W, Zhu G et al (2014) A review of the preparation and application of flavour and essential oils microcapsules based on complex coacervation technology. J Sci Food Agric 94:1482–1494
13. Lamprecht A, Schäfer U, Leh CM (2001) Influences of process parameters on preparation of microparticle used as a carrier system for O - 3 unsaturated fatty acid ethyl esters used in supplementary nutrition. J Microencapsul 18:347–357

Chapter 10

Microencapsulation of Oils and Lipophilic Bioactive Compounds in Multilayer O/W Emulsions

Pedro Renann Lopes de França and Louise Emy Kurozawa

Abstract

Proteins are widely employed as emulsifiers; however, emulsions derived from them can exhibit limited resistance to environmental stresses, which compromises their stability and the protection of encapsulated active compounds. The formation of electrostatic complexes between proteins and polysaccharides at the oil-in-water O/W emulsion interface leads to multilayer emulsions, which offer enhanced characteristics and provide a viable alternative to traditional methods by overcoming these limitations. Therefore, this chapter presents the protocol of the Layer-by-Layer method for the preparation of multilayer emulsions through the manipulation of the charge properties of proteins and polysaccharides, aiming to form protein-polysaccharide complexes at the emulsion interface.

Keywords Protein, Polysaccharide, Interface, Charge, Layer-by-layer

1 Introduction

Technological advancements in the industry require the development of new emulsifiers and methodologies for producing stable emulsions, with the aim of creating new products or enhancing their shelf life, ensuring efficient delivery, and maximizing the efficacy of active ingredients. Oil-in-water (O/W) emulsions have been widely used in the cleaning, cosmetics, pharmaceutical, and food industries to encapsulate various bioactive compounds, improving their solubility and stability [1].

However, conventional O/W emulsions may exhibit several limitations, such as restricted stability across pH ranges, salt concentrations, heating, dehydration, freezing, and cooling [2]. Such disadvantages drive research to identify alternative preparation methods to enhance emulsion stability. One strategy involves creating covalent protein-polysaccharide complexes with favorable surface activity, which improves the performance of various emulsified systems [3, 4]. This approach results in multilayer emulsifiers at the O/W interface

Ana Silvia Prata and Louise Emy Kurozawa (eds.), *Biopolymers*, Methods and Protocols in Food Science,
https://doi.org/10.1007/978-1-0716-5230-5_10,

with enhanced characteristics, offering a viable alternative to address these limitations.

Multilayer emulsions are obtained through a process known as layer-by-layer (LBL), with the electrostatic deposition of hydrocolloids with opposite charges at the O/W interface [5]. Strong electrostatic complexes can form between charged proteins and polysaccharides with opposite charges. This not only provides the previously mentioned benefits but also ensures increased oxidative protection of the encapsulated oils [6] and improved digestibility [7].

Multilayer emulsions can be formed using proteins below their isoelectric point (with a net positive charge) and anionic polysaccharides [5, 8], or proteins above their isoelectric point (with a net negative charge) and cationic polysaccharides [9]. Additionally, it is possible to alternate between anionic and cationic polysaccharides to create systems with more than two interfacial layers [10].

Thus, this chapter aims to present a practical method for obtaining multilayer O/W emulsions using the LBL technique, combining proteins with charged polysaccharides. It also provides a simple method for determining the concentration of polysaccharides to be used and for observing the final emulsions.

2 Material

2.1 Reagents

1. Hydrophobic core material: fixed (or non-volatile) oil [5], diacylglycerol [11], essential oils [12], and pigments [13].
2. Protein: whey, soy, pea proteins, etc. (*see* **Note 1**).
3. Polysaccharide: pectin, alginate, chitosan, etc. (*see* **Note 2**).
4. Acetone.
5. Ethanol (≥99.9%).
6. Fluorescent dyes: Calcofluor White, Fluorescein isothiocyanate-5 (FITC), and Nile Red.
7. Fuming hydrochloric acid (37%).
8. Sodium hydroxide.
9. Prepare all solutions using ultrapure water (conductivity = 18 MΩ·cm at 25°C) and use only reagent-grade chemicals

2.2 Laboratory Glassware

1. 100 mL graduated cylinder.
2. 1000 mL, 100 mL, 50 mL, and 10 mL volumetric flasks.
3. Automated pipettes (100 μL, 200 μL, and 1000 μL).
4. Glass beakers.
5. Microscopy slides and coverslips.
6. 1.5 mL microtubes.
7. Pasteur pipettes.

2.3 Equipments

1. Analytical balance (0.01 mg precision).
2. Fluorescence microscope or laser scanning confocal microscope.
3. Magnetic stirrer and magnetic bar.
4. pH meter.
5. Rotor-stator homogenizer.
6. Ultrasonic bath.
7. Vortex mixer.
8. Zetasizer microelectrophoresis chamber and cuvettes for zeta potential measurement.

3 Methods

Figure 1 presents a practical schematic of the preparation of multilayer emulsions.

3.1 Determination of pH for Emulsion Preparation

Before producing multilayer emulsions, it is essential to determine the pH of the suspensions to be adopted for their preparation. This can be achieved by measuring the zeta potential of the suspensions at different pH levels (3–10), as described in the protocol of complex coacervation. Some appointments are made during the:

1. Adjustment of the pH of the solutions (*see* **Notes 3 and 4**).
2. The definition of the complexation pH (*see* **Notes 5 and 6**).

3.2 Determination of Protein Concentration

An important step is determining the protein concentration to be used. This required concentration can vary based on several factors, including the nature of the protein, the amount of lipophilic compound in the emulsion, and the system's viscosity [14]. Consequently, the optimal concentration is specific to each system under study (*see* **Note 7**).

Additionally, the process might include evaluating emulsion stability to determine the minimum amount of protein necessary to prevent phase separation over time. Important methods for assessing emulsion stability include:

- Techniques for evaluating emulsion stability, such as Turbiscan emulsion stability analysis [5] or the creaming index [15].
- Measurement of droplet size in the emulsion through static light scattering using a laser diffraction device [5], which can also monitor changes in droplet size over time.
- Observation of the formed emulsion using optical microscopy [16], which can provide valuable information about coalescence and flocculation within the emulsion.

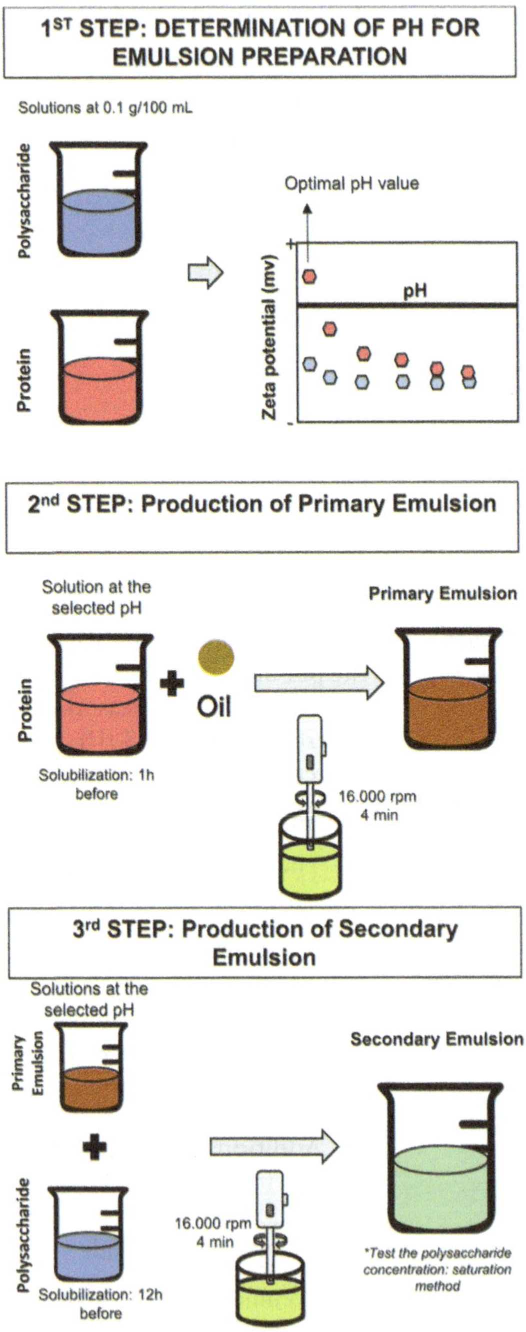

Fig. 1 General scheme for the microencapsulation of lipophilic compounds in multilayer emulsions

Additional methods for characterizing emulsion stability and observing destabilization processes by various techniques are detailed in McClements [17].

3.3 Determination of Polysaccharide Concentration to be Associated with Protein by Interfacial Saturation Method for Droplets

The concentration of polysaccharide to be used in conjunction with the protein will be specific to each system, as it must account for various factors, such as the desired viscosity of the emulsion, protein concentration, droplet size, etc. However, it is possible to determine the minimum concentration necessary for the formation of multilayer emulsions through the evaluation of the zeta potential of the emulsions using a method known as the saturation method [2].

1. Prepare emulsions with varying concentrations of polysaccharide (*see* **Note 8**) (see Sect. 3.4).
2. Using a pipette, collect 100 µL from each emulsion and dilute it in 19.9 mL of ultrapure water.
3. Immediately measure the zeta potential using a Zetasizer microelectrophoresis chamber, as described previously in Sect. 3.1.

The minimum concentration required is determined when the charge inversion of the emulsion occurs between the primary and secondary emulsions, and this value becomes approximately constant (saturation point) (Fig. 2). This concentration is considered necessary for the formation of interfacial multilayers along the dispersed droplets in the emulsion [1].

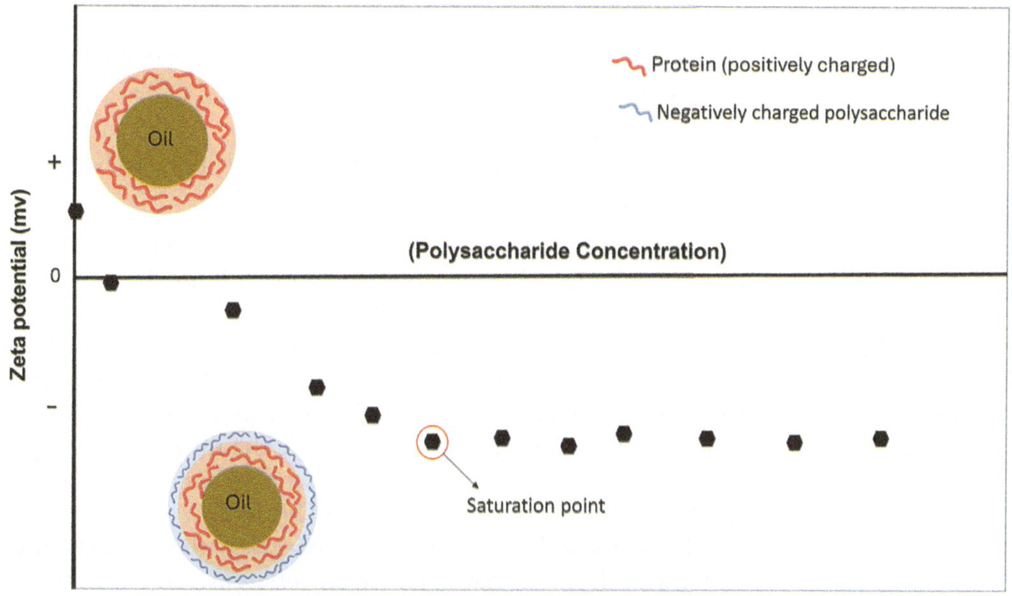

Fig. 2 Schematic representation of the zeta potential evaluation of multilayer emulsions formed using the association of a protein below its isoelectric point and an anionic polysaccharide. The saturation point indicates the ideal concentration for the formation of interfacial multilayers

3.4 Production of Emulsions

The layer-by-layer method involves the production of consecutive emulsions, with each layer of emulsifiers being added to form interfacial layers.

3.4.1 Preparation of Protein and Carbohydrate Solutions

Solution A: Protein Suspension

1. Weigh the required amount of protein and add it to water.
2. Maintain agitation for approximately 1 h at ambient temperature.

Solution B: Carbohydrate Solution

1. Weigh the desired amount of carbohydrate to be used in association with the protein.
2. Carefully add the carbohydrate to water. Maintain agitation using a magnetic stirrer for 12 h at ambient temperature (*see* **Note 9**).

All solutions should be adjusted to the pH using HCl solution (1 M) or NaOH solution (1 M).

3.4.2 Primary Emulsion

1. Before starting the production of the emulsions, check the pH of all solutions and adjust if necessary.
2. Add the lipophilic compound of interest to the protein suspension (Solution A). Homogenize the system using a rotor-stator homogenizer at 16,000 rpm for 4 min.

3.4.3 Secondary Emulsion

1. Combine the primary emulsion with carbohydrate solution (Solution B) in a 1:1 ratio.
2. Keep under gentle magnetic stirring for approximately 10 minutes to ensure system homogenization.
3. Verify the pH of the emulsion to ensure proper electrostatic interactions between the protein and polysaccharide (*see* **Note 10**).

The process can be successively repeated to form tertiary, quaternary, or additional emulsions (*see* **Notes 11 and 12**).

3.5 Verification of Multilayer Emulsion Formation

Evaluation of charge inversion and droplet saturation, as described in Sect. 3.3 can provide strong indications of multilayer emulsion formation with interfacial coating through electrostatic deposition. However, it is recommended to observe the formation of electrostatic layers using fluorescence microscopy techniques [5] or laser scanning confocal microscopy [9] (*see* **Note 13**).

To perform these methods, it is necessary to stain the components with dyes that have affinity for different components of the multilayer emulsion. For a bilayer system (composed of protein, oil, and carbohydrate), observations can be made using the following dyes: Calcofluor White (carbohydrate), Nile Red (oil), and FITC (protein) [18].

The detailed information can be found in Chap. 17.

Consult the specific manual for each microscope used for processing micrographs. It is recommended to observe each component separately and overlay the obtained micrographs to draw important conclusions about the interactions between the different constituents (*see* **Notes 14, 15, and 16**).

4 Notes

1. Proteins with reasonable solubility are essential for stabilizing these systems, similar to other non-pickering emulsified systems [5]. Furthermore, the proteins must exhibit proven interfacial activity to effectively stabilize the emulsified system [19].
2. For formulating multilayer emulsions, the polysaccharide used must be a polysaccharide with a net charge. This charge can be negative, as in pectin [5] or alginate [20], or positive, as in chitosan [21]. Neutral polysaccharides should not be considered, as they do not interact with proteins.
3. **pH Range for Zeta Potential**: Generally, a pH range of 3–10 is sufficient for most systems, as recommended by the manufacturer of the Zetasizer Nano Series cuvettes. However, if necessary, lower pH values, such as pH 2, can be evaluated. Determining zeta potential at such low pH may be challenging and could degrade the cuvettes, rendering them unusable.
4. **Polysaccharide Solubility**: Be aware of the solubility of the polysaccharide used. For example, chitosan has limited solubility at alkaline pH levels.
5. **Isoelectric Point Consideration**: The isoelectric point (IP) of the protein is a key parameter. Typically, emulsions are obtained below the IP, with the protein carrying a positive charge and the Polysaccharide being anionic (e.g., pectin, alginate). It is important to maintain a pH reasonably far from the IP, as approaching the IP can reduce the functional properties of proteins [7].
6. **Polysaccharide Solubilization:** The required solubilization pH depends on the specific carbohydrate used. For instance, chitosan is soluble only in acidic pH. Thus, every experiment will also be constrained by the solubility range of the polysaccharide.
7. We recommend consulting the following method: "Preparation of Dietary Lipid Emulsions Using Laboratory Methods" [22], in which the authors present approaches for emulsion stabilization and for determining the minimum

protein concentration in the continuous phase required to fully cover the droplet population. This enables subsequent stabilization of the emulsion based on droplet size after homogenization.

8. **Polysaccharide Concentrations**: Concentrations can vary significantly. Generally, concentrations ranging from 0.005% to 1% w/w are sufficient for effective multilayer formation.
9. **Bubble Formation**: Solutions of pectin, alginate, chitosan, and similar polysaccharides may form bubbles during homogenization, which is undesirable for emulsion formation. An ultrasonic bath for 2 min can be used to remove these bubbles.
10. **Neutral Polysaccharide**: When using neutral polysaccharides such as maltodextrin, they can be prepared in conjunction with charged polysaccharides, as neutral polysaccharides do not participate in multilayer formation.
11. **Active Compound Dilution**: Keep in mind that the active compound will become increasingly diluted, and the emulsions may be less concentrated relative to it.
12. **Homogenization Methods**: The method suggested here uses a rotor-stator homogenizer. However, these systems may not always produce emulsions with adequate droplet sizes. Additional homogenization methods, such as high-pressure homogenization, ultrasonic treatment, or microfluidization, may be used to achieve kinetically more stable emulsions [23].
13. **Microscopy Techniques**: Both fluorescence and confocal laser scanning microscopy can be used to assess multilayer formation. However, confocal microscopy provides higher resolution and more detailed images, allowing for more precise conclusions about the evaluated systems.
14. **Dye Stability**: Generally, if stored correctly, dye solutions have a long shelf life. However, it is recommended to test the fluorescence capacity of the dye on a fluorescence or confocal microscope before use.
15. **Dyeing Components**: If observing all components is challenging, consider staining the individual components before preparing the emulsion. Perform microscopy shortly after preparation.
16. **Droplet Visualization**: If oil droplets are too close together for clear visualization, it may be necessary to dilute the sample with water in a 1:1 ratio.

References

1. Burgos-Díaz C, Wandersleben T, Marqués AM et al (2016) Multilayer emulsions stabilized by vegetable proteins and polysaccharides. Curr Opin Colloid Interface Sci 25:51–57
2. Guzey D, McClements DJ (2006) Formation, stability and properties of multilayer emulsions for application in the food industry. Adv Colloid Interface Sci 128–130:227–248
3. Dickinson E (2003) Hydrocolloids at interfaces and the influence on the properties of dispersed systems. Food Hydrocoll 17:25–39
4. Schmitt C, Sanchez C, Desobry-Banon S et al (1998) Structure and technofunctional properties of protein-polysaccharide complexes: a review. Crit Rev Food Sci Nutr 38:689–753
5. D França PRL, Gomes MHG, Nascimento RF et al (2024) Microencapsulation of linseed oil in multilayer emulsions stabilized by carob protein hydrolysate and pectin. Food Hydrocoll 150:109641
6. Tamm F, Härter C, Brodkorb A et al (2016) Functional and antioxidant properties of whey protein hydrolysate/pectin complexes in emulsions and spray-dried microcapsules. LWT 73:524–527
7. Zeeb B, Lopez-Pena CL, Weiss J et al (2015) Controlling lipid digestion using enzyme-induced crosslinking of biopolymer interfacial layers in multilayer emulsions. Food Hydrocoll 46:125–133
8. Fioramonti SA, Martinez MJ, Pilosof AMR et al (2015) Multilayer emulsions as a strategy for linseed oil microencapsulation: effect of pH and alginate concentration. Food Hydrocoll 43:8–17
9. Ding M, Liu L, Zhang T et al (2021) Effect of interfacial layer number on the storage stability and in vitro digestion of fish oil-loaded multilayer emulsions consisting of gelatin particle and polysaccharides. Food Chem 336:127686
10. Shi Y, Liu Y, Sun Y et al (2024) Interfacial multilayer self-assembly of protein and polysaccharides: ultrasonic regulation, stability and application in delivery lutein. Int J Biol Macromol 272:132880
11. Diao X, Jia R, Wang Y et al (2024) The physicochemical properties, microstructure, and stability of diacylglycerol-loaded multilayer emulsion based on protein and polysaccharides. LWT 196:115879
12. Tadele W, Kulawik P, Stępień A et al (2024) Effects of multilayer nano/mini furcellaran/chitosan emulsions with oregano essential oil and bioactive peptides on sensory, physicochemical properties and retrogradation in sushi in cold storage conditions. Innov Food Sci Emerg Technol 96:103767
13. Sun X, Liu C, Wang D et al (2023) Effects of coating layers chitosan/pectin on lipid stability and in vitro digestion of astaxanthin-loaded multilayer emulsions. LWT 173:114282
14. Maphosa Y, Jideani VA (2018) Factors Affecting the Stability of Emulsions Stabilised by Biopolymers. In: Karakuş S (ed) Science and Technology Behind Nanoemulsions. InTech
15. Gomes MHG, Kurozawa LE (2020) Improvement of the functional and antioxidant properties of rice protein by enzymatic hydrolysis for the microencapsulation of linseed oil. J Food Eng 267:109761
16. Hu Y-T, Ting Y, Hu J-Y et al (2017) Techniques and methods to study functional characteristics of emulsion systems. J Food Drug Anal 25:16–26
17. Mcclements DJ (2007) Critical review of techniques and methodologies for characterization of emulsion stability. Crit Rev Food Sci Nutr 47:611–649
18. Partanen R, Sibakov J, Rommi K et al (2016) Dispersion stability of non-refined turnip rapeseed (Brassica rapa) protein concentrate: impact of thermal, mechanical and enzymatic treatments. Food Bioprod Process 99:29–37
19. Li X, Wu G, Qi X et al (2019) Physicochemical properties of stable multilayer nanoemulsion prepared via the spontaneously-ordered adsorption of short and long chains. Food Chem 274:620–628
20. Shi F, Chang Y, Shen J et al (2023) A comparative investigation of anionic polysaccharides (sulfated fucan, ι-carrageenan, κ-carrageenan, and alginate) on the fabrication, stability, rheology, and digestion of multilayer emulsion. Food Hydrocoll 134:108081
21. Xu D, Aihemaiti Z, Cao Y et al (2016) Physicochemical stability, microrheological properties and microstructure of lutein emulsions stabilized by multilayer membranes consisting of whey protein isolate, flaxseed gum and chitosan. Food Chem 202:156–164
22. Beaumal V, Marze S (2024) Preparation of dietary lipid emulsions using laboratory methods. In: Lopez C, Genot C, Riaublanc A (eds) Multidimensional characterization of dietary lipids. Springer US, New York, NY, pp 289–299
23. Swarbrick J (ed) (2013) Homogenization and Homogenizers. In: Encyclopedia of pharmaceutical science and technology. 4th edn. CRC Press, pp 1848–1854

Check for updates

Chapter 11

Pickering Emulsion and Encapsulation

Francisco E.R. Paes and Carolina S.F. Picone

Abstract

Pickering emulsion (PE), or emulsion stabilized by particles, is a promising vehicle for encapsulation of bioactive compounds as its stability is usually superior to traditional emulsions. Depending on the characteristics of the stabilizing particles is possible to produce water-in-oil or oil-in-water emulsions, enabling the encapsulation of hydrophilic or hydrophobic compounds. PEs can be easily produced by low-cost and flexible equipment, being an interesting alternative to traditional emulsified systems for encapsulation. This chapter presents the methodology for encapsulation of hydrophobic compounds by oil-in-water Pickering Emulsions. The main parameters to be considered for a successful PE encapsulation are discussed.

Keywords Particles, Biopolymers, Bioactive, Stability

1 Introduction

Pickering emulsions (PEs) are dispersed systems composed of two immiscible liquids, typically oil and water, in which one is dispersed into the other as droplets. Unlike conventional emulsions, which rely on surfactants that continuously adsorb and desorb at the interface, PEs are stabilized by solid particles that strongly anchor at the oil-water interface. These particles form a viscoelastic mechanical barrier that prevents droplet coalescence, ensuring stability through steric repulsion [1]. The effectiveness of this stabilization depends on the particles' contact angle, which determines their ability to anchor to the interface. This stabilization mechanism differs from conventional emulsions in which the interface is stabilized by low molecular weight surfactants [2]. Usually, PE presents longer kinetic and temperature stability than conventional emulsions, and enables the use of natural particles instead of synthetic surfactants for interface stabilization, favoring the development of more healthy and sustainable food carriers. In addition, PE might present enhanced controlled digestion and sustainable release of the bioactive compared to regular emulsions [3].

Ana Silvia Prata and Louise Emy Kurozawa (eds.), *Biopolymers*, Methods and Protocols in Food Science,
https://doi.org/10.1007/978-1-0716-5230-5_11,

The characteristics and successful stability of PEs strongly depend on how the particles anchor onto the interface, which is affected by particle properties such as wettability, morphology, size, and their hydrophilic-hydrophobic balance. The hydrophilic or hydrophobic properties of the particles are related to the particle's three-phase contact angle (θ) and determine the type of emulsion to be formed [4]. Particles with $\theta < 90°$ have great affinity for the aqueous phase and are prone to stabilize oil-in-water (O/W) PEs. Water-in-oil (W/O) PEs are formed by particles with $\theta > 90°$ due to the hydrophobicity of the particles, which are preferentially wetted by the oil phase [5]. When at the interface, the particles acquire a high desorption energy that needs to be overcome for the particle to be released from the interface. This high energy is responsible for the greater stability of PEs in comparison with regular emulsions, in which the emulsifier diffuses and spontaneously adsorbs and desorbs from the interface [6].

The shape of particles affects the desorption energy, with anisotropic particles presenting much higher desorption energy, and thus being more prone to form stable PEs. The decrease in particle's size and contact angle tends to reduce its desorption energy and facilitate particle detaching from the interface, resulting in less stable PEs. Generally, a 1:10 ratio is often observed between the particle size and PE droplet size [6].

To assure the interface stabilization, the amount of particles must be enough to recover the interfacial area to be stabilized. Assuming that all particles in the emulsion are located at the interface, the mass of particles needed to stabilize droplets with diameter d_d can be calculated by Eq. 1 [6]:

$$m_p = \frac{4\rho_p d_p V_d C}{d_d} \qquad (1)$$

where m_p is the mass of particles (kg), ρ_p is the particle density (kg/m^3), V_d is the volume of the dispersed phase (m^3), and C is the degree of coverage.

Currently, particles of different sizes and morphologies have been used to stabilize PEs as aggregated proteins (such as whey [7], zein [8], and soy proteins [9]), rod-like polysaccharides as cellulose nanocrystals [10], spheric polysaccharides such as starches [11, 12], and fibrillar particles as nanocellulose [13] and nanochitin [14]. The diversity of particles that can be used on PE stabilization enables the incorporation of the bioactive compounds not only within the inner phase of the emulsion but also within the particles themselves, making PE an outstanding system for bioactive encapsulation [15].

The next section presents the methodology of encapsulation of a hydrophobic bioactive in O/W Pickering Emulsion. Details about the application of this technique to the encapsulation of quercetin are described in the notes [16].

2 Materials

1. Aqueous phase containing stabilizing particles (*see* **Note 1**).
2. Oil phase loaded with the bioactive compound (*see* **Note 2**).
3. Glass vials 40 mm × 21 mm with cap.
4. High-speed homogenizer.
5. Pipettes.

3 Methods

In order to encapsulate a hydrophobic bioactive by O/W PE it is necessary to previously prepare the aqueous phase containing the particles (contact angle < 90°), and the oil phase loaded with the target bioactive compound [17]. Then, the proper amounts of each phase must be homogenized. Different types of equipment can be used, such as high-speed homogenizer [14], ultrasound [18], high-pressure homogenizer [19], and high-pressure microfluidizer [20]. Herein, we describe the PE formation by high-speed homogenizer, which is a widely used equipment on emulsification process and can provide enough energy for the formation of PEs (Fig. 1).

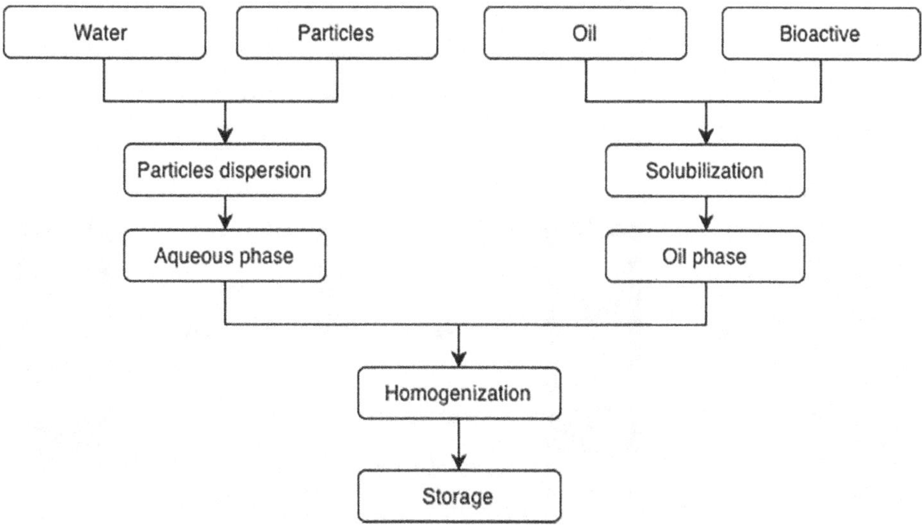

Fig. 1 Flowchart of encapsulation of a hydrophobic bioactive in O/W PE

3.1 Encapsulation by PE

The encapsulation of hydrophobic bioactive on O/W PE (Fig. 2) will be performed as illustrated in Fig. 2.

1. Homogenize the aqueous phase containing the particles and the oil phase loaded with the bioactive compound of interest (1:2 v/v) (*see* **Notes 1 and 2**) in a high-speed homogenizer for 2 min at 13,500 rpm at room temperature (*see* **Note 3**) ensuring a homogeneous flow throughout the sample.
2. Close the flask and store the resulting PE at room temperature.
3. Store under light protection to prolong the shelf life of the system.

3.2 Validation of PE Formation

Verify if an opaque and viscous liquid is obtained (Fig. 2d). In a different scenario, the presence of two phases (Fig. 2b) after the formation of PE indicates the instability of PE (*see* **Notes 4 and 5**).

Note x: The stability of PE can be evaluated by visual changes in PE over time. Changes in PE color can be indicative of phase separation. The presence of sediments on the bottom of the flask or an oil layer on the top can also be related to emulsion instability. A simple methodology with adaptation is presented below to assess the success of emulsion formation and its stability [21].

1. Store the PE for 7 days at room temperature.
2. Periodically evaluate the appearance of the emulsions concerning changes in their color, settling (accumulation of particles on the bottom of the flask), phase separation, creaming (accumulation of droplets on the top of the flask), or the presence of oily phase on the top (*see* **Note 4**).
3. In case of phase separation, measure the height of the creamed layer (Hc) and the total phase (Ht) with a rule (see Note 15).
4. Calculate the Creaming Index (CI).

$$\mathrm{CI}(\%) = \frac{H_c}{H_t} * 100 \tag{2}$$

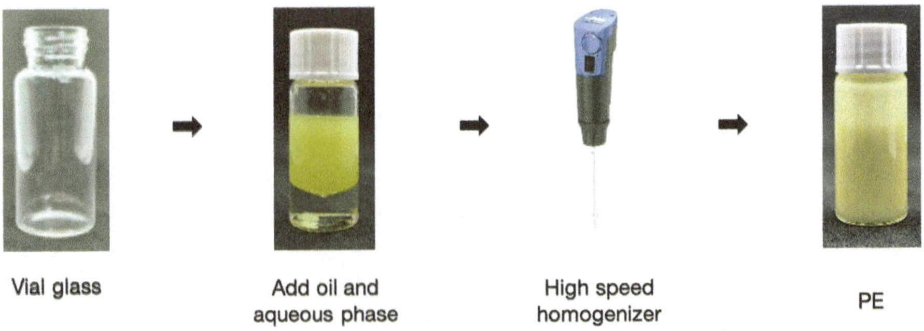

Fig. 2 Schematic experimental setup. (**A**) Empty vial glass; (**B**) Vial with oil and aqueous phase; (**C**) High speed homogenizer; (**D**) Freshly prepared PE

4 Notes

1. To encapsulate the hydrophobic bioactive (e.g., quercetin) on PE stabilized by chitosan particles, it is suggested the production of aqueous phase with 0.1% (w/v) chitosan particles. To do this, measure 0.2 g of low molecular weight chitosan using an analytical balance and place it in a 200 mL beaker. Next, suspend the 0.2 g of chitosan in 200 mL of ultrapure water and stir the solution for 2 min at 25°C. Adjust the pH of the chitosan solution to pH 4.1 with 30 M acetic acid for complete solubilization. Keep the chitosan solution under magnetic stirring for 24 h at 25°C. Then, adjust the pH of the chitosan solution to 6.7 (with 1 M sodium hydroxide) under stirring at 25°C to form the chitosan particles.
2. Weigh 0.1 g of quercetin into a 150 mL beaker and add 100 mL of sunflower oil, stir until complete solubilization. In case of light-sensitive compounds, the encapsulation under the protection of light is recommended. The contact angle and stability of the PE depend on the composition and polarity of the oil used. We recommend using sunflower oil due to its low viscosity and favorable emulsification properties. Other oils can be used but may show different results.

    1. In a glass vial measuring 40 mm × 21 mm, add 3 mL of the aqueous phase and 6 mL of the oil phase with the assistance of pipettes. The glass vial is an important part of production of PE; it must be clear, and the dimensions of the glass flasks should ensure homogeneous flow of the solution during the homogenization step and prevent areas of fluid stagnation.
    2. Microscopy is a good tool to verify the success of the emulsification process. It can also be used to check the stability of the emulsion over time. Kinetic stable PE emulsions are expected to maintain their morphology during storage, keeping the same droplet size and aggregation state. Cryo-SEM can be used to observe particle arrangement at the droplet interface, while confocal microscopy helps determine whether the bioactive compound is encapsulated within the droplets. However, these techniques require prior sample preparation and advanced equipment. Conversely, optical microscopy is a simple and traditional technique that can confirm successful bioactive emulsification and assess system stability over time [22].

3. Figure 3 shows the morphology of fresh prepared PE by optical microscopy.

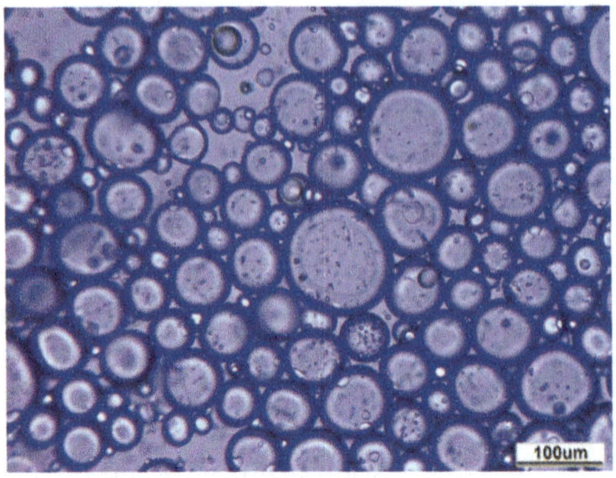

Fig. 3 Optical microscopy of quercetin encapsulated in O/W PE (just after preparation)

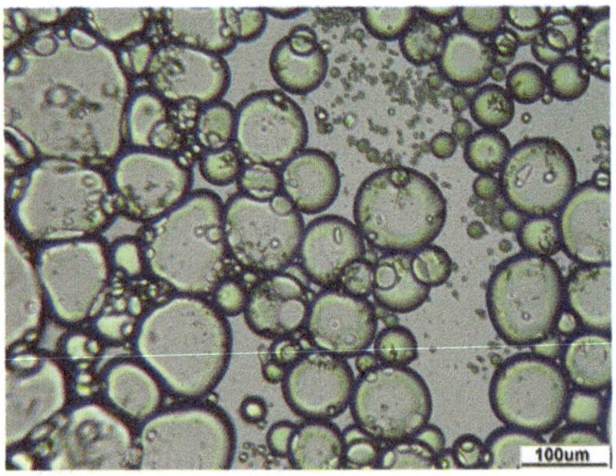

Fig. 4 Optical microscopy of quercetin encapsulated in O/W PE (after 5 months)

4. Figure 4 shows the morphology of quercetin encapsulated in O/W PE after 5 months of storage at 25°C. The increase in droplet's size in comparison with Fig. 4 is an indicative of droplets coalescence.
5. Do the PE at room temperature, as high temperatures can alter the desorption energy of the particles and the rheological properties of the phases, promoting destabilization.
6. Make sure that all equipment and glassware are clean and free from surfactant and alcohol residues. Impurities can affect the results.
7. Properly adjust the homogenizer stirring rod to ensure optimal emulsion flow and avoid air incorporation.
8. Seal the emulsion container tightly to avoid evaporation.
9. Store in a dry and clean environment.

10. Check the quality of the oil used; oils with impurities or that are oxidized can affect the bioactive's stability.
11. Consider the length and polarity of the fatty acids present in the oil used; long-chain fatty acids are more difficult to emulsify than short and medium-chain fatty acids.
12. It is important to use the phases completely solubilized. Phases with aggregates of uncertain size can alterPE stability.
13. The concentrations and the proportion between the aqueous and oil phases described in this chapter are specific to this methodology. Using different particles, continuous phases, and dispersed phases may require adjustments in the methodology.
14. The pH and ionic strength are important factors for Pickering emulsions stability. They can affect the surface charge of particles and possible electrostatic interactions between the droplets. Therefore, the pH and ionic strength of the aqueous phase can be used as a tool to inhibit coalescence and postpone PE destabilization and bioactive release [22].

References

1. Ni L, Yu C, Wei Q, Liu D, Qiu J (2022) Pickering emulsion catalysis: interfacial chemistry, catalyst design, challenges, and perspectives. Angew Chem 134:202115885
2. Low LE, Siva SP, Ho YK, Chan ES, Tey BT (2020) Recent advances of characterization techniques for the formation, physical properties and stability of Pickering emulsion. Adv Colloid Interface Sci 277:102117
3. Cheng C, Yuan C, Cui B, Li J, Liu G (2024) β-Cyclodextrin based Pickering emulsions for α-tocopherol delivery: antioxidation stability and bioaccessibility. Food Chem 438:138000
4. Heidari F, Jafari SM, Ziaiifar AM (2023) Surface modification of silica nanoparticles by chitosan for stabilization of water-in-oil Pickering emulsions. Carbohydr Polym Technol Appl 6:100381
5. Sarkar A, Dickinson E (2020) Sustainable food-grade Pickering emulsions stabilized by plant-based particles. Curr Opin Colloid Interface Sci 49:69–81
6. Schroën K, Shen X, Hasyyati FI, Deshpande S, Gucht JVD (2024) From theoretical aspects to practical food Pickering emulsions: formation, stabilization, and complexities linked to the use of colloidal food particles. Adv Colloid Interface Sci 334:103321
7. Zhou L, Liu Y, Li Y, Zhou W, Peng S, Qin X, Liu G, Zhang H (2024) Pickering emulsion stabilized by glycosylated whey protein isolate complexed with chitooligosaccharide for the improving stability of delivery and bioaccessibility of DHA. Food Hydrocoll 151:109858
8. Cui S, McClements DJ, He X, Xu X, Tan F, Yang D, Sun Q, Dai L (2024) Interfacial properties and structure of Pickering emulsions co-stabilized by different charge emulsifiers and zein nanoparticles. Food Hydrocoll 146:109285
9. Xu YT, Liu TX, Tang CH (2019) Novel pickering high internal phase emulsion gels stabilized solely by soy β-conglycinin. Food Hydrocoll 88:21–30
10. Bai L, Lv S, Xiang W, Huan S, McClements DJ, Rojas OJ (2019) Oil-in-water Pickering emulsions via microfluidization with cellulose nanocrystals: 1. formation and stability. Food Hydrocoll 96:699–708
11. Wang PP, Luo ZG, Chen C, Fu X, Tamer TM (2020) Effects of octenyl succinic anhydride groups distribution on the storage and shear stability of Pickering emulsions formulated by modified rice starch. Carbohydr Polym 228:115389
12. Saari H, Wahlgren M, Rayner M, Sjöö M, Matos M (2019) A comparison of emulsion stability for different OSA-modified waxy maize emulsifiers: granules, dissolved starch, and non-solvent precipitates. PLoS One 14:e0210690
13. Chevalier RC, Júnior FDO, Cunha RL (2024) Modulating digestibility and stability of Pickering emulsions based on cellulose nanofibers. Food Res Int 178:113963

14. Zhu Y, Huan S, Bai L, Ketola A, Shi X, Zhang X, Ketoja AJ, Rojas OJ (2020) High internal phase oil-in-water pickering emulsions stabilized by chitin nanofibrils: 3D structuring and solid foam. ACS Appl Mater Interfaces 12:11240–11251
15. Noor N, Gani A, Jhan F, Shah MA, ul Ashraf Z (2022) Ferulic acid loaded pickering emulsions stabilized by resistant starch nanoparticles using ultrasonication: characterization, in vitro release and nutraceutical potential. Ultrason Sonochem 84:105967
16. Liu H, Wang C, Zou S, Wei Z, Tong Z (2021) Simple, reversible emulsion system switched by pH on the basis of chitosan without any hydrophobic modification. Langmuir 28:11017–11024
17. Nikfarjam N, Qazvini NT, Deng Y (2015) Surfactant free Pickering emulsion polymerization of styrene in w/o/w system using cellulose nanofibrils. Eur Polym J 64:179–188
18. Apostolidis E, Gerogianni A, Anagnostaki E, Paximada P, Mandala I (2024) Assembly of spherical-shaped resistant starch nanoparticles to the oil droplet surface promotes the formation of stable oil in water Pickering emulsions. Food Hydrocoll 151:109775
19. Zhang S, Holmes M, Ettelaie R, Sarkar A (2020) Pea protein microgel particles as Pickering stabilisers of oil-in-water emulsions: responsiveness to pH and ionic strength. Food Hydrocoll 102:105583
20. Zhu Y, McClements DJ, Zhou W, Peng S, Zhou L, Zou L, Liu W (2020) Influence of ionic strength and thermal pretreatment on the freeze-thaw stability of Pickering emulsion gels. Food Chem 303:125401
21. Yang H, Su X, Meng X, Kennedy JF, Liu B (2020) Fabrication and characterization of Pickering emulsion stabilized by soy protein isolate-chitosan nanoparticles. Food Hydrocoll 247:116712
22. Xiao J, Wang XA, Gonzalez AJP, Huang Q (2016) Kafirin nanoparticles-stabilized Pickering emulsions: microstructure and rheological behavior. Carbohydr Polym 54:30–39

Check for updates

Chapter 12

Nanoprecipitation

Ana Silvia Prata

Abstract

Nanoprecipitation is a straightforward and reproducible technique widely employed for the preparation of polymeric nanoparticles intended for the encapsulation of bioactive compounds. This chapter describes a standardized protocol for the nanoprecipitation of starch, a biodegradable, renewable, and biocompatible polysaccharide, highlighting its advantages as a carrier matrix for food, pharmaceutical, and cosmetic applications. The method involves the controlled precipitation of starch by using an antisolvent under constant stirring, yielding nanoparticles with tunable size, stability, and encapsulation efficiency. Overall, this protocol offers a versatile approach for designing starch-based nanocarriers, facilitating the protection, delivery, and controlled release of sensitive active ingredients.

Keywords Starch, Antisolvent, Nanoparticles, Controlled release, Biodegradable carriers

1 Introduction

Nanoprecipitation, a bottom-up technique based on a thermodynamic process of controlled molecular assembly, involves the precipitation of a solute (polymer) from a solution, normally water, using an antisolvent, in which the polymer is insoluble [1]. The requirement of the antisolvent is the water-miscibility, and the most common are acetone, ethanol, and isopropanol. The precipitation can occur after a gradual addition of the dilute polymer solution to the nonsolvent or successive addition of the nonsolvent into the polymer solution [2], as shown in Fig. 1. The formation mechanism implies the interphase turbulence, diffusion, and surface interactions at the nonequilibrium stage, and the obtention of nanoparticles depends on processing conditions, the composition of both phases and the physicochemical interactions between phases. Biodegradable or non-biodegradable polymers have been used for the preparation of nanoparticles intended for in vivo studies [3]. Particularly when considering the precipitation of starches, the process begins with their dissolution in water, and then an antisolvent is introduced, which creates a supersaturated environment, causing the solute to

Ana Silvia Prata and Louise Emy Kurozawa (eds.), *Biopolymers*, Methods and Protocols in Food Science,
https://doi.org/10.1007/978-1-0716-5230-5_12,

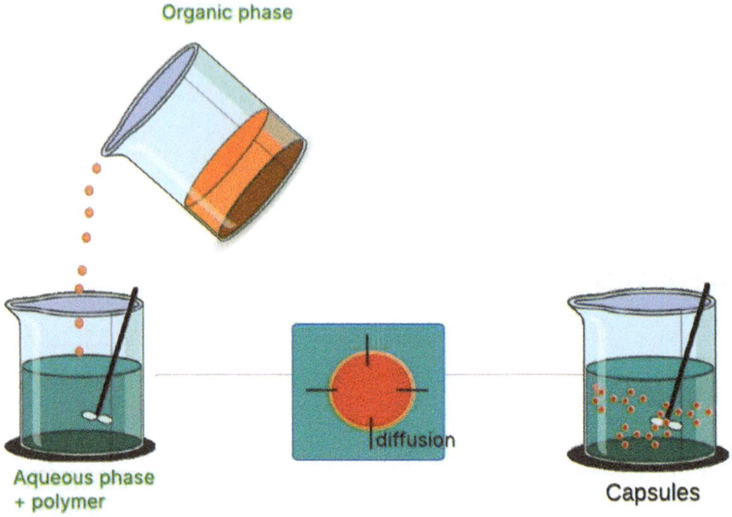

Fig. 1 Schema of the nanoprecipitation process

precipitate out of the solution. Obtained particles could be either nanospheres or nanocapsules, if an oily phase is added [3].

To prevent these nanoparticles from clustering together, stabilizers or surfactants are added. These molecules adsorb onto the nanoparticles' surface, providing a barrier against aggregation. Centrifugation is used to collect the nanoparticles, which are subsequently rinsed with water, solvent, or their mixture to remove the surfactant. The antisolvent evaporates, hardening the nanoparticles, and then it is preferred to adopt organic solvents with low boiling points, such as ethanol, acetone, hexane, or methylene chloride [4].

Nanoprecipitation offers good control over particle size by adjusting polymer concentration and mixing ratios, and provides several benefits, including simplicity and the absence of complex equipment or operation conditions. However, encapsulation efficiency for hydrophillic agents may be limited, and scaling up to produce large quantities can be challenging.

2 Materials

- Normal corn starch (25%–26.5% amylose content) (*see* **Notes 1, 2, and 3**).
- Lutein: lipophilic active compound (*see* **Note 4**).
- Ethanol (*see* **Note 5**).
- Water.
- Sodium tripolyphosphate (STPP).

- Syringe and needle.
- Indium standard (for calibration).
- Aluminum DSC pans (sample + reference).
- Magnetic stirrer or an ultrasonic bath for stirring or emulsifying the reaction.
- Water bath for controlling the reaction temperature.
- Centrifuge for particle collection.
- Freeze dryer.
- DSC instrument.
- X-ray diffractometer.
- Drying oven.
- Analytical balance.
- FT-IR spectrometer.

3 Preparation of Nanoparticles

1. Dissolve the 1.5 g of starch powder in 50 mL of distilled water (3% w/v of starch solution) using a magnetic stirrer at 1000–2000 rpm (*see* **Notes 6, 7, and 8**).
2. Place the starch solution in a water bath at 90°C, and keep continuously stirring at 700 rpm for 30 min. Then the mixture is removed from the heat source and allowed to cool down to about 60°C (*see* **Note 9**).
3. Dissolve 90 mg of the poorly soluble core material in 100 mL of ethanol (*see* **Notes 10 and 11**).
4. Optional step - Crosslinking: This procedure may be employed when the active compound is mixed with the starch solution. Add 0.5 g of STPP into 10 mL distilled water and then add gently to the starch solution (10:90 STTP:starch-loaded core material v/v). Keep the reaction mixture at room temperature for 2 h with constant agitation rate at 25°C (*see* **Note 12**).
5. For nanoprecipitation, add the ethanolic solution dropwise to the dispersed starch at a constant speed (500 rpm) for 1 h. The addition starts when the starch solution is yet warm (*see* **Notes 13 and 14**).
6. Wash the resultant precipitate with 80/20 absolute ethanol/water and centrifuge to remove unreacted compounds. Repeat once again this procedure (*see* **Note 15**).
7. Finally, perform a washing step with absolute ethanol, and isolate the nanoparticles by centrifugation for 1 h at 4500 rpm (*see* **Note 16**).

8. At the end, take this supernatant for further analysis to determine the loss of core material.
9. Freeze at −80°C the resultant nanoparticles and freeze-dry them at a pressure of 0.2 mbar for 24 h.

4 Characterization

4.1 Differential Scanning Calorimetry (DSC)

- Weigh 5 mg of freeze-dried nanoparticles into an aluminum DSC pan.
- Add 10 mg of distilled water to the pan and mix gently.
- Seal the DSC pan.
- Allow the sealed sample to equilibrate overnight at room temperature.
- Turn on the DSC and allow it to stabilize.
- Perform temperature calibration using indium.
- Place an empty aluminum pan in the reference position.
- Load the prepared sample pan into the DSC.
- Set the temperature program:
- Initial temperature: 30°C.
- Final temperature: 130°C.
- Heating rate: 10°C/min.
- Start the run and allow the instrument to record thermograms.
- From the DSC thermogram, determine: Onset temperature (To), Peak temperature (Tp), Conclusion temperature (Tc), Enthalpy change (ΔH).

4.2 X-Ray Diffraction (XRD)

- Dry both native starch and SNP samples at 40°C for 24 h before analysis.
- Set scanning parameters: 2θ range: 6–50°, Radiation: Cu Kα, Voltage: 40 kV, Current: 30 mA.
- Record diffractograms for all samples.
- Determine crystal type and relative crystallinity.

4.3 Fourier Transform Infrared Spectroscopy (FT-IR)

Set spectral range to 4000–400 cm^{-1}.
Set resolution to 4 cm^{-1}.
Configure the instrument to perform 32 scans per sample.

Place the sample in the spectrometer.
Perform scanning across the defined range.
Record FT-IR spectra.
Identify characteristic absorption peaks related to starch structure.
The particle should also be evaluated for size and encapsulation efficiency as described in Chap. 17.

5 Notes

1. Other starchy materials can also be employed such as potato starch (20%–28.0% amylose content), sweet potato starch (20.6% amylose content) and tapioca starch (18.9% amylose content), waxy corn starch (0.8% amylose content), pea starch (35%–40.0% amylose content) and high amylose corn starch (69.0% amylose content). Other polymers can also be employed [5].
2. Parameters like starch sources, starch concentration, the volume ratio of starch solution and solvent, and surfactant employed influence the nanoprecipitation and nanoparticles characteristics [1, 5].
3. Amylose is known to form some inclusion complexes, depending on the core material; higher amylose content can improve the retention.
4. Hydrophilic drugs have rapid migration leading to drug loss in the aqueous phase. Modifications may be studied to increase the retention [6]. Commonly poorly soluble core materials encapsulated are Piroxicam, Paclitaxel, lutein, curcumin, and diclofenac sodium.
5. The most used solvents in nanoprecipitation method are ethanol, propanol, isopropanol, butanol, acetone, hexane, methylene chloride, or dioxane. Butanol can only form a complex and precipitate with amylose, but not with amylopectin [1].
6. Some authors dissolve the starch in acetone [6].
7. The aqueous phase, in addition to distilled water, may also contain other chemical compounds as sodium hydroxide, glycerin, urea, and DMSO. The sodium hydroxide breaks the hydrogen bonds of the starch, and urea or glycerin prevents the self-association of the starch molecules. A ratio employed for the aqueous-solvent system of NaOH, glycerin (or urea), and distilled water is 0.8:1:98.2 w/w. Only water can also be adopted [7].
8. The starch concentration may vary from 0.5%–5% w/v.

9. The gelatinization of the starch may vary according to the source. The incubation occurs at 80–150°C for 30 min with constant stirring (700 rpm). The temperature of 80–100°C is proposed for waxy starch, 100°C for starches with 18%–28% amylose, and 150°C for high amylose content starches (~70%) [8].
10. The core material can also be solubilized in the water phase with a surfactant solution (2% tween 80). The ratio surfactant:starch is 20:70 (v/v) [7].
11. To provide nanocapsules instead of nanospheres, mineral oil or vegetable oil would be added [3].
12. Crosslinkers can enhance the granule stability with new covalent bonds, which allow for increasing the desired stability conditions. Normally, the greater the amount of STPP (0.5, 1, and 2 g), the greater the particle sizes [7].
13. The ratio aqueous solution:organic varies between 1:1 to 1:4 v/v [7, 8].
14. O/W surfactant is regularly used to avoid clustering. The surfactant solution is included in the organic phase [8, 9].
15. Other solutions may be used in this step, such as pure organic phase but also, centrifugation, and resuspension in water.
16. Other centrifugation conditions can be adopted, such as 7500 rpm for 5 min; 5000 rpm for 10 min, 3000 rpm for 15 min

References

1. Marta H, Rizki DI, Mardawati E, Djali M, Mohammad M, Cahyana Y (2023) Starch nanoparticles: preparation, properties and applications. Polymers (Basel) 15(5):1167
2. Morán D, Gutiérrez G, Blanco-López MC, Marefati A, Rayner M, Matos M (2021) Synthesis of starch nanoparticles and their applications for bioactive compound encapsulation. Appl Sci 11(10):4547
3. Rivas CJM, Tarhini M, Badri W, Miladi K, Hélène Greige-Gerges H, Nazari QQ, Rodríguez SAG, Román RA, Fessi H, Elaissari A (2017) Nanoprecipitation process: from encapsulation to drug delivery. Int J Pharm 532(1):66–81. https://doi.org/10.1016/j.ijpharm.2017.08.064
4. Pulingam T, Foroozandeh P, Chuah JA, Sudesh K (2022) Exploring various techniques for the chemical and biological synthesis of polymeric nanoparticles. Nanomaterials 12(3):576
5. Hernández-Giottonini KY, Quiñones-Rabago JA, Peñuñuri-Miranda O, Rodríguez-Córdova RJ, Zavala-Rivera P, Lucero-Acuña A (2024) Starch nanoparticle preparation by the nanoprecipitation technique: effects of formulation parameters. Colloids Surf A Physicochem Eng Asp 702:135022
6. Acevedo-Guevara L, Nieto-Suaza L, Sanchez LT, Pinzon MI, Villa CC (2018) Development of native and modified banana starch nanoparticles as vehicles for curcumin. Int J Biol Macromol 111:498–504
7. El-Naggar ME, El-Rafie MH, El-sheikh MA, El-Feky GS, Hebeish A (2015) Synthesis, characterization, release kinetics and toxicity profile of drug-loaded starch nanoparticles. Int J Biol Macromol 81:718–729
8. Chen Y, Ding L, Di H, Kain Kirkensgaard JJ, Khakimov B, Sun B et al (2024) Green preparation of small-sized starch nanoparticles using nanoprecipitation. Food Hydrocoll 153:109974
9. Moorthi C, Kathiresan K (2013) Fabrication of highly stable sonication assisted curcumin nanocrystals by nanoprecipitation method. Drug Invent Today 5(1):66–69

Chapter 13

Pressurized CO_2 Antisolvent Coprecipitation of Bioactive Compounds in Polymeric Materials

Julia T. Paula, Ana P. Da Fonseca Machado and Ana C. Aguiar

Abstract

Coprecipitation of bioactive compounds in polymeric materials offers key advantages such as enhanced stability, bioavailability, and controlled release, making it valuable for pharmaceutical and nutraceutical formulations. Pressurized and supercritical fluids, especially carbon dioxide (CO_2), are promising alternatives for coprecipitation due to their nontoxic, environmentally friendly, and cost-effective nature, with mild critical properties that prevent thermal degradation. Many techniques use CO_2 as an antisolvent to precipitate bioactives and polymers in a single step, controlling particle properties through process parameters, such as the supercritical antisolvent (SAS) technique. However, challenges include phase equilibrium issues, nozzle clogging, and scale-up difficulties. Thus, the aim of this chapter is to outline a reliable step-by-step procedure for coprecipitating low-solubility bioactive compounds and polymers in pressurized CO_2 by the SAS technique.

Keywords Encapsulation, Pressurized fluids, Supercritical fluids, Supercritical antisolvent, Carbon dioxide

1 Introduction

Coprecipitation of bioactive compounds in polymeric materials offers several advantages, including improved stability, bioavailability, and controlled release. Thus, the coprecipitation processes are valuable for developing effective pharmaceuticals, nutraceuticals, and ingredients for functional foods. In the last decades, sub- and supercritical technologies have been applied as sustainable alternatives for encapsulation/coprecipitation of bioactive compounds [1–3] and pharmaceutical drugs [4–6].

The use of pressurized and supercritical fluids, especially carbon dioxide (CO_2), offers a promising alternative to conventional encapsulation and coprecipitation processes. CO_2 is nontoxic, nonflammable, environmentally acceptable, relatively low cost, and leaves no toxic residues in the processed material. Additionally, its mild critical properties (critical temperature of 31.1 °C and critical pressure of

Ana Silvia Prata and Louise Emy Kurozawa (eds.), *Biopolymers*, Methods and Protocols in Food Science,
https://doi.org/10.1007/978-1-0716-5230-5_13,

73.8 bar) help reduce operating costs and prevent thermal degradation of thermosensitive bioactive compounds [7, 8].

Techniques utilizing supercritical and pressurized fluids are classified according to the role of the pressurized fluid: solvent, solute, or antisolvent [9]. The processes that use pressurized CO_2 as an antisolvent enable coprecipitation of the bioactive compound and polymer material in a single step; unlike those that use pressurized CO_2 as a solute or as a solvent. In the antisolvent precipitation, the compound of interest and the polymer material are dissolved in an organic solvent, and this solution is then atomized through a nozzle along with subcritical or supercritical CO_2 under controlled temperature and pressure conditions [3].

The principle of antisolvent techniques is based on differences in solubility of the polymer in the organic phase and in the antisolvent phase. Additionally, the organic solvent exhibits a stronger affinity for the antisolvent than for the solute, and when mixed together, the antisolvent reduces the solvation power of the organic solvent, causing the solute to precipitate [10].

Small changes in process parameters, such as pressure, temperature, bioactive compound-to-encapsulating agent ratio, CO_2 flow rate, solution flow rate, nozzle size, etc., allow the control of size distribution and morphology of particles [11].

The main disadvantages associated with coprecipitation processes using pressurized CO_2 as an antisolvent are related to the difficulty in controlling factors that influence the process such as flow rate and solution concentration, the use of organic solvents, limitations arising from phase equilibrium between different system components (CO_2, extract/bioactive compound, solvent, and polymer), operational issues such as the clogging of nozzles and pipelines, and challenges related to process scale-up [12].

The objective of this chapter is to present a reliable experimental step-by-step procedure for producing coprecipitates of bioactive compounds with low solubility in pressurized CO_2 in polyvinylpyrrolidone (PVP) using the pressurized CO_2-assisted coprecipitation technique, where CO_2 acts as an antisolvent.

2 Materials

- Beakers.
- Stainless steel spatula.
- Petri plates or other utensils to help collect particles.
- Amber storage vessels.
- Dry and milled vegetable material to be extracted or high-purity bioactive compound.
- Qualitative filter paper.

- Absolute ethanol (99.5% purity).
- Polyvinylpyrrolidone (PVP) polymer (average molecular weight: 10,000 g/mol).
- Carbon dioxide (CO_2: 99.8% purity).
- Silica.
- Analytical balance (precision of 10 mg).
- Ultrasonic bath with fixed frequency (50 kHz) and power (50 W).
- Vacuum filtration system.
- High-pressure precipitation unit (Fig. 1) composed of a CO_2 tank, high-pressure pump for liquid CO_2 and solution, CO_2 cooling system (cooler), stainless-steel precipitation covered by an electric heating jacket, injection system with an internal diameter of 100 μm coupled to the top cover of the precipitation vessel, a stainless-steel filter at the bottom of the precipitation vessel, an automated back-pressure regulator, and stainless-steel jacketed cyclone separator.

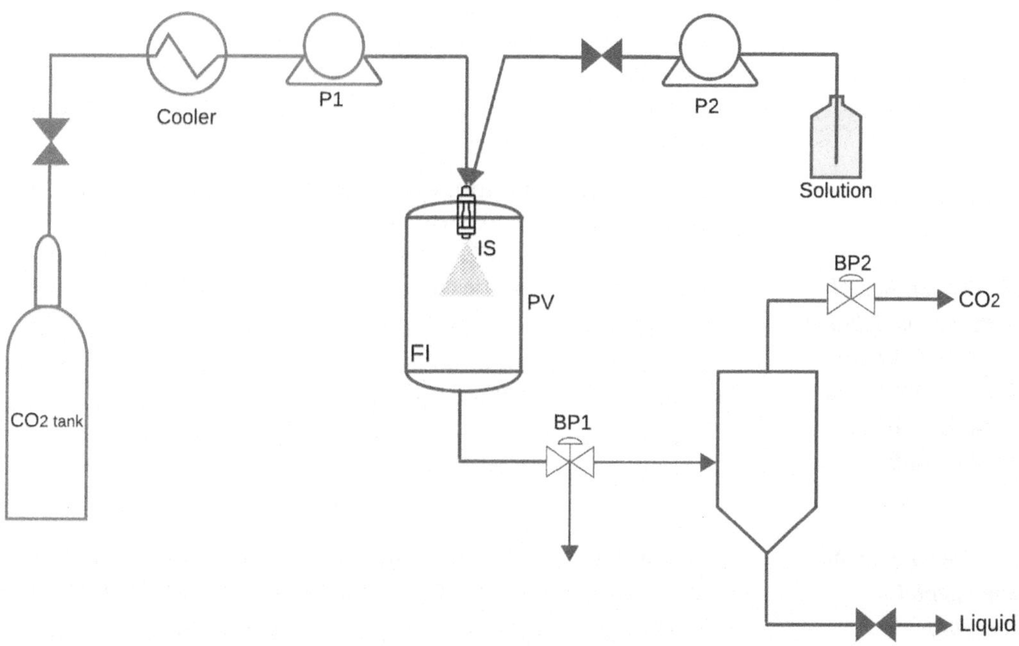

Fig. 1 Schematic diagram of a general precipitation unit. *P1* CO_2 pump, *P2* solution, *IS* injection system, *PV* precipitator vessel, *FI* filter, *BP1 and BP2* back-pressure regulators

3 Methods

3.1 Extraction of Bioactive Compounds

If the material to be coprecipitated is obtained from a plant matrix, the extraction step must initially be carried out. A general extraction process for coprecipitation of polar compounds (due to their low solubility in pressurized CO_2) is described below (*see* **Notes 1–4**).

1. Add 200 mL of ethanol to a 250 mL beaker containing approximately 35 g of the dry and milled sample to be extracted.
2. Manually macerate the solid–liquid mixture until it reaches a homogeneous appearance.
3. Place the macerated mixture in an ultrasonic bath for 15 min. Monitor the temperature increase during the process—if the temperature of the mixture increases during sonication, use a bath with cold water (8°C) recirculation or, if possible, reduce the ultrasound power of the equipment.
4. After the sonication step, vacuum filter the solid–liquid mixture and recover the filtrated extract in an amber storage vessel.
5. Store the vessel under refrigeration until the extract is used in coprecipitation experiments.

3.2 Preparation of the Bioactive Compound Solution

If the high-purity bioactive compound to be coprecipitated is in powder or dry form, it must be solubilized as described below(*see* **Notes 4–6**).

1. Weigh the material using a spatula and a beaker on an analytical balance.
2. Add the organic solvent ethanol and homogenize the system until complete dissolution.
3. Store the container under refrigeration until the extract is used in coprecipitation experiments.

3.3 Preparation of the Polymeric Solution (Ethanolic Extract/Bioactive Compound-Ethanolic Solution + PVP) (see Notes 5, 7, and 8)

1. Weigh the PVP in a beaker using a spatula and analytical balance.
2. Completely dissolve the PVP in the ethanolic extract/bioactive compound-ethanolic solution at a dry-basis ratio of 0.8:1.0 (PVP:dry extract/bioactive compound).

3.4 Coprecipitation Experimental Procedure

1. Begin the coprecipitation system preparation by turning on the CO_2 cooling system, the solution pump, and the electric heaters.
2. Check the nozzles and piping to inspect for leaks and blockages.
3. Purge the liquid pump with solvent ethanol to remove any residual solution and air bubbles.

4. Pump the solvent ethanol through the equipment's piping to the precipitation vessel inlet to remove any residual solution and air bubbles.
5. After checking for leaks, purging, and pumping the solutions, close the precipitation vessel and adjust the valves to the appropriate positions to start the coprecipitation process.
6. Once the cooling system and heating system have reached the appropriate temperatures, open the CO_2 tank and pump CO_2 into the precipitation vessel until the desired pressure is reached (*see* **Notes 9 and 10**).
7. Pump the solvent ethanol for a period of 2–5 min into the precipitation vessel, while maintaining a constant addition of pressurized CO_2 (antisolvent), in order for the system to reach steady-state conditions.
8. Stop pumping the pure solvent and then pump the polymeric solution into the precipitation chamber (*see* **Notes 11 and 12**) for a defined period of time, also maintaining the constant addition of the pressurized antisolvent (CO_2 in the supercritical or subcritical state, depending on the pressure and temperature conditions).
9. After introducing the predetermined amount of solution (polymer and extract/bioactive compound), which will vary depending on the desired mass of dry particles, stop the addition and close the solution inlet valve on the pressure vessel.
10. After completing the solution introduction, continue the flow of antisolvent for a sufficient time to remove residual ethanol present in the generated particles/coprecipitates—about 50 min for 75 mL of solution (*see* **Notes 13 and 14**).
11. System depressurization: after the drying period, stop the CO_2 flow and slowly depressurize the system to atmospheric pressure (*see* **Note 15**).
12. After the system is fully depressurized, carefully open the top cover of the precipitation vessel. Once opened, recover the coprecipitated material from both the inner wall and the filter attached to the bottom of the precipitation vessel in a Petri dish or other utensils that you consider practical.
13. Weigh the coprecipitated material using an analytical balance and then transfer it to a sealed flask containing silica to prevent the particles from absorbing moisture from the environment.
14. After collection, store the recovered particles at low temperatures (<8°C) to prevent decomposition before analysis.
15. Once the coprecipitation process is complete, clean the unit using absolute ethanol or 70% v/v ethanol to prevent any clogging in the pipes and accessories (*see* **Note 16**).

4 Notes

1. To prepare the extract, it is recommended to study the influence of the solvent and the extraction method on the composition of the extract obtained. Other techniques such as maceration, pressurized liquid extraction, microwave-assisted extraction, etc. can be used—more information can be obtained in the literature [13].
2. It is recommended to prepare a single batch of extract in order to avoid differences in the final composition of the particles due to different batches of injected extracts.
3. When preparing extracts, other systems can be used to separate the solid material from the extract in addition to the vacuum filtration system, such as separation by centrifugation.
4. The selection of the solvent is critical. High-purity solvents are essential for preparing solutions to dissolve the polymer and extract/bioactive compounds, as this helps prevent the presence of water or other insoluble compounds in pressurized CO_2. While dimethyl sulfoxide, N-methyl-2-pyrrolidone, dichloromethane, acetone, and ethanol are commonly used, ethanol is the preferred solvent for food and pharmaceutical applications due to its GRAS (Generally Recognized as Safe) status [14]. The general recommendation is to avoid or limit the use of other solvents whenever possible.
5. In order to calculate the appropriate dilution considering the ratio between the mass of the polymeric material and the dry extract, it is necessary to determine the total solids content of the extract to be coprecipitated, as well as the concentration of the bioactive compound of interest in the extract.
6. Defining the concentration of the bioactive extract/solution is usually the key to the success of coprecipitation using pressurized antisolvent techniques. Very dilute extracts/solutions do not lead to the formation of dry and fine homogeneous powders, but rather of a sticky material. Therefore, on a laboratory scale, concentrations of extracts/bioactive solutions above 15–20 mg/mL are recommended.
7. The choice of polymeric material depends on the desired release rate and the nature of the target compound. Hydrophilic polymers such as polyvinylpyrrolidone (PVP) and Pluronic F127 are commonly used for poorly water-soluble compounds such as liposoluble vitamins [15, 16] or phenolic compounds [17]. For hydrophobic agents, polylactic acid (PLA) and poly(L-lactic acid) (PLLA) are typical choices, often used for delivering anticancer drugs [18, 19].

8. Extract/bioactive compound:polymer ratios above 1:2.5 usually lead to the production of fully agglomerated particles. Therefore, high concentrations of PVP should be avoided.
9. The control of particle size distribution and morphology is highly dependent on process parameters such as pressure, temperature, bioactive compound to encapsulating agent ratio, CO_2 flow rate, solution flow rate, and nozzle size [11]. Fine-tuning these parameters is essential for achieving the desired particle characteristics.
10. The selection of experimental conditions for the precipitation process must be done carefully so that it can be conducted successfully. It is useful to know the phase equilibrium of the system to define the experimental conditions. Relevant information can be found in the literature [17].
11. Before preparing the particles, ensure the extract solution has reached room temperature and inspect it for suspended solids. If any are present, ensure the solute is fully dissolved. If the solution is saturated with the presence of precipitated material, repeat the filtration step to prevent clogging of pipes and nozzles.
12. The flow rate of the feed solution must be adjusted so as not to overload the system. Therefore, generally lower flow rates are required (1.0–3.0 mL/min, for commercial laboratory scale unit) for successful coprecipitation.
13. Extra attention must be paid to the particle drying stage—this step is extremely important, since the amount of organic solvent remaining in the medium can cause the particles to coalesce and lose their original characteristics and is a recurring problem in coprecipitation processes of bioactive compounds and polymers in pressurized CO_2.
14. For aqueous solutions, antisolvent precipitation processes are not suitable for coprecipitation due to the low water solubility in pressurized CO_2.
15. The depressurization step following particle drying is critical. The process must be conducted at a slow rate to prevent CO_2 flow from altering the particles' shape or causing aggregation.
16. The equipment cleaning step is essential for the success of the coprecipitation process. The equipment must be completely cleaned with absolute ethanol or 70% v/v ethanol in order to avoid incrustations in the pipes and accessories, thus leaving it in suitable conditions for the next experimental run.

References

1. Quintana SE, Villanueva-Bermejo D, Reglero G, García-Risco MR, Fornari T (2019) Supercritical antisolvent particle precipitation and fractionation of rosemary (*Rosmarinus officinalis* L.) extracts. J CO2 Util 34:479–489
2. Machado APF, Rezende CA, Rodrigues RA, Barbero GF, Rosa PTV, Martínez J (2018) Encapsulation of anthocyanin-rich extract from blackberry residues by spray-drying, freeze-drying and supercritical antisolvent. Powder Technol 340:553–562
3. Arango-Ruiz Á, Martin Á, Cocero MJ, Jiménez C, Londoño J (2018) Encapsulation of curcumin using supercritical antisolvent (SAS) technology to improve its stability and solubility in water. Food Chem 258:156–163
4. Vallejo R, Quinteros D, Gutiérrez J, Martínez S, Rodríguez Rojo S, Ignacio Tártara L, Palma S, Arias FJ (2024) Acetazolamide encapsulation in elastin like recombinamers using a supercritical antisolvent (SAS) process for glaucoma treatment. Int J Pharm 657:124098
5. Remiro PFR, Rosa PTV, Moraes ÂM (2022) Effect of process variables on imiquimod micronization using a supercritical antisolvent (SAS) precipitation technique. J Supercrit Fluids 181:105500.
6. Alias D, Yunus R, Chong GH, Abdullah CAC (2017) Single step encapsulation process of tamoxifen in biodegradable polymer using supercritical anti-solvent (SAS) process. Powder Technol 309:89–94
7. Cocero MJ, Martín A, Mattea F, Varona S (2009) Encapsulation and co-precipitation processes with supercritical fluids: fundamentals and applications. J Supercrit Fluids 47:546–555
8. Martín A, Cocero MJ (2008) Micronization processes with supercritical fluids: fundamentals and mechanisms. Adv Drug Deliv Rev 60:339–350
9. Martinez J, de Aguiar AC, Machado APF, Barrales FM, Vigano J, Santos P (2020) Process integration and intensification. In: Cifuentes A (Ed.) Comprehensive foodomics. Elsevier, Amsterdam, Netherlands, 786–807
10. Sang-Do Y, Kiran E (2005) Formation of polymer particles with supercritical fluids: areview. J Supercrit Fluids 34:287–308
11. Temelli F (2018) Perspectives on the use of supercritical particle formation technologies for food ingredients. J Supercrit Fluids 134:244–251
12. Nascimento Junior DR, Tabernero A, Albuquerque ECDMC, Vieira de Melo SAB (2021) Biopesticide encapsulation using supercritical CO_2: a comprehensive review and potential applications. Molecules 26:4003
13. Panja P (2018) Green extraction methods of food polyphenols from vegetable materials. Curr Opin Food Sci 23:173–182
14. Paola P, De Marco I (2020) Supercritical antisolvent process for pharmaceutical applications: a review. Processes 8:938
15. Prosapio V, Reverchon E, De Marco I (2017) Incorporation of liposoluble vitamins within PVP microparticles using supercritical antisolvent precipitation. J CO2 Util 19:230–237
16. Adeli E (2016) The use of supercritical anti-solvent (SAS) technique for preparation of Irbesartan-Pluronic® F-127 nanoparticles to improve the drug dissolution. Powder Technol 298:65–72
17. Machado APF, Montes A, Valor D, Fernández-Ponce MT, Barbero GF, Maróstica Júnior MR, Pereyra C, de la Ossa EM (2022) Co-precipitation of grape residue extract using sub- and supercritical CO_2 technology. J CO2 Util 61:102010
18. Franco P, De Marco I (2020) Supercritical antisolvent process for pharmaceutical applications: a review. Processes 8:938
19. Wischke C, Schwendeman SP (2008) Principles of encapsulating hydrophobic drugs in PLA/PLGA microparticles. Int J Pharm 364:298–327

Chapter 14

Bioactive Compounds Vehiculated by Oleogels

Matheus A.S. Roman and Rosiane L. Cunha

Abstract

Oleogels have been used as potential source of lipid structuring in food matrices, aiming at replacing saturated fats. In addition to this application, these structured oily systems can vehiculate lipophilic bioactive compounds. Different methodologies can be used to obtain oleogels, and this chapter presents a sequence of protocols for obtaining these three-dimensional structures for carrying bioactive molecules.

Keywords Direct approach, Indirect approach, Emulsion-template, Foam-template, Oxidation, Lipid structuring

1 Introduction

Oleogels are semisolid materials composed of oil physically entrapped in a three-dimensional (3D) gel network formed by structuring agents or oleogelators. Different structuring agents are used to form stable oily structures that can be used as partial or total replacements for saturated and trans fats in processed foods [1]. There are two categories of structuring agents used to form oleogels: low molecular weight gelling agents (LMWGs), such as waxes, fatty acids, mono- and diglycerides, sterols, and lecithin, and high molecular weight gelling agents (HMWGs) represented by biopolymers, such as proteins, starches, and cellulose derivatives.

The first step to obtain an oleogel is to establish its application and then identify a suitable oleogelator depending on the chosen process. The method of producing oleogels can be classified into two categories called direct and indirect approaches [2–5]. The first is commonly used to produce oleogels with oleogelators from the LMWG group, consisting of dissolving the structuring agent in the oil under heating oil and subsequent cooling to form the structure. In contrast, the second approach is related to structuring agents from the HMWG group. The oleogels obtained by this last approach require more processing steps, such as the prior formation

Ana Silvia Prata and Louise Emy Kurozawa (eds.), *Biopolymers*, Methods and Protocols in Food Science,
https://doi.org/10.1007/978-1-0716-5230-5_14,

of an emulsion (oleogels via emulsion-template) or a foam (oleogels via foam-template).

Due to their three-dimensional structure and hydrophobic nature, oleogels are competitive as matrices for the delivery of lipophilic bioactive compounds, which are mostly unstable during food processing or even under physiological conditions [6]. Oleogels can thus prevent the loss of functionality of the bioactive molecule but, at the same time, can help preserve the oil, rich in unsaturated fatty acids, against lipid oxidation [5].

An important issue that must be considered when vehiculating bioactive compounds into oleogels is the possibility of changing their three-dimensional structure due to the interaction between the structuring agent and the bioactive molecule. Oleogels obtained by direct approach are produced with oleogelators that have affinity with lipophilic bioactive compounds, which can alter, for example, the crystallization of monoglycerides [7]. This fact can prevent the formation of a sufficiently stable crystalline network capable of maintaining the oil and the bioactive molecules incorporated into the matrix.

On the other hand, in oleogels obtained by indirect approach, mainly those based on emulsion, the emulsifier and the bioactive compound can compete for the oil droplet interface, weakening the interfacial film that prevents flocculation and/or coalescence of the droplets. Stable emulsions are essential for the production of oleogels by emulsion-template. Exploring different proportions of emulsifiers and bioactive compounds is important to address the associated challenge.

This chapter aims to help young researchers develop oleogels using the main approaches established in the literature to incorporate lipophilic bioactive compounds into these structures. The division between oleogels obtained by direct and indirect approaches was carried out, followed by the main strategies for vehiculating bioactive molecules into the respective oleogels.

2 Materials

1. Oleogel by direct approach
   - Oleogelator or structuring agent (e.g., waxes, mono- and diglycerides, ethylcellulose, sterols such as γ-oryzanol and β-sitosterol, and lecithin) with known melting points.
   - Food-grade oil (e.g., soybean, sunflower, corn, or olive oil, medium-chain triglycerides).
   - Beaker (10, 50, 100 mL—the volume depends on the total volume of oil+oleogelator).

- Heating plate with magnetic stirrer.
- Magnetic bar.
- Cooling system (e.g., ice bath, water bath, or incubation chamber).
- Temperature-controlled oven.

2. Oleogel by indirect approach

- Food-grade oil (e.g., soybean, sunflower, corn, or olive oil, medium-chain triglycerides).
- Distilled water, ultrapure water, or buffer solution (if necessary).
- Emulsifier (e.g., proteins, food-grade synthetic surfactants, polymers).
- Heating plate with magnetic stirrer.
- Magnetic bar.
- High shear mixer (*see* **Note 1**).
- Petri dish (internal diameter approximately 130 mm).
- Spatula.
- Aluminum pans.
- Forced-circulation oven (if necessary).
- Ultra-freezer or liquid nitrogen.
- Freeze dryer.
- Hydrophobic bioactive compound (e.g., phenolic compounds, vitamins).
- Jacketed beaker.
- Circulation thermostatic bath.

3 Methods

3.1 Oleogels by Direct Approach

Oleogels may deliver bioactive compounds in their crystalline network (described in Sect. 3.1.2) and emulsions or nanoemulsions that use the oleogel as the lipid fraction (described in Sect. 3.1.3).

3.1.1 Determining the Composition of the Oleogel

1. In a glass flask (e.g., beaker), mix the structuring agent with the oil at a temperature above the melting point of the oleogelator, and keep it under magnetic stirring until the oleogelator is completely dissolved (*see* **Notes 2 and 3**).

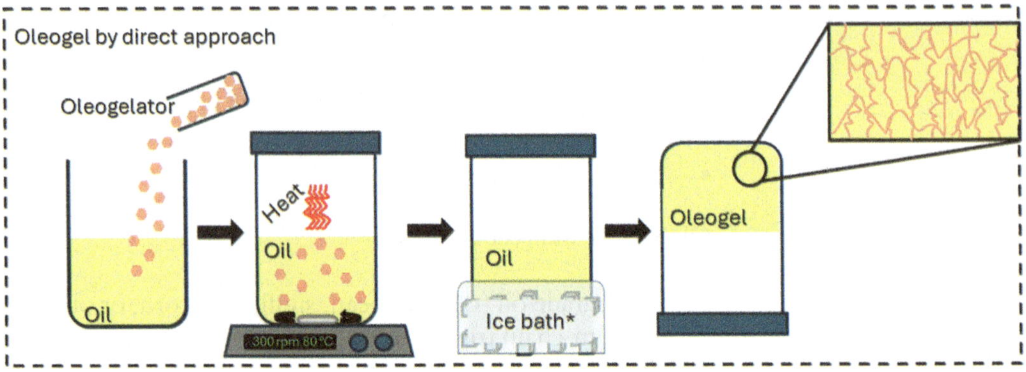

Fig. 1 Schematic steps to obtain the oleogel by direct approach. *The ice bath can be replaced by another cooling system, such as a water bath or an incubation chamber

2. Immediately transfer the flask with the mixture (oil+oleogelator) to a cooling system and let it cool for at least 30 min or until solidification.
3. Turn the flask upside down to see if the oleogel has formed (tilt test). If it sticks to the bottom and sides of the flask, the structure has managed to trap all the oil. If the mixture runs down the sides, it may be necessary to increase the amount of oleogelator and return to Step 1.
4. After cooling, the oleogel already formed must be stored at a constant temperature for 24 h before being characterized (*see* **Note 4**). Figure 1 shows the steps of the process.

3.1.2 Bioactive Compound-Loaded Oleogels

1. Follow Step 1 described in Sect. 3.1.1 in the preselected concentration of oil and oleogelator.
2. Add the lipophilic bioactive compound and an emulsifier to the mixture containing the oil and the oleogelators. Keep the mixture under magnetic stirring until complete solubilization, maintaining a constant temperature above the melting temperature of the oleogelator (*see* **Note 5**).
3. Follow Steps 2–4 described in Sect. 3.1.1, shown in the upper part of Fig. 2.

3.1.3 Bioactive Compounds in Emulsion-Loaded Oleogels

1. Follow Step 1 described in Sect. 3.1.1.
2. Add the lipophilic bioactive compound to the mixture containing the oil and the oleogelators. Keep the mixture under magnetic stirring until complete solubilization, maintaining a constant temperature above the melting temperature of the oleogelator.

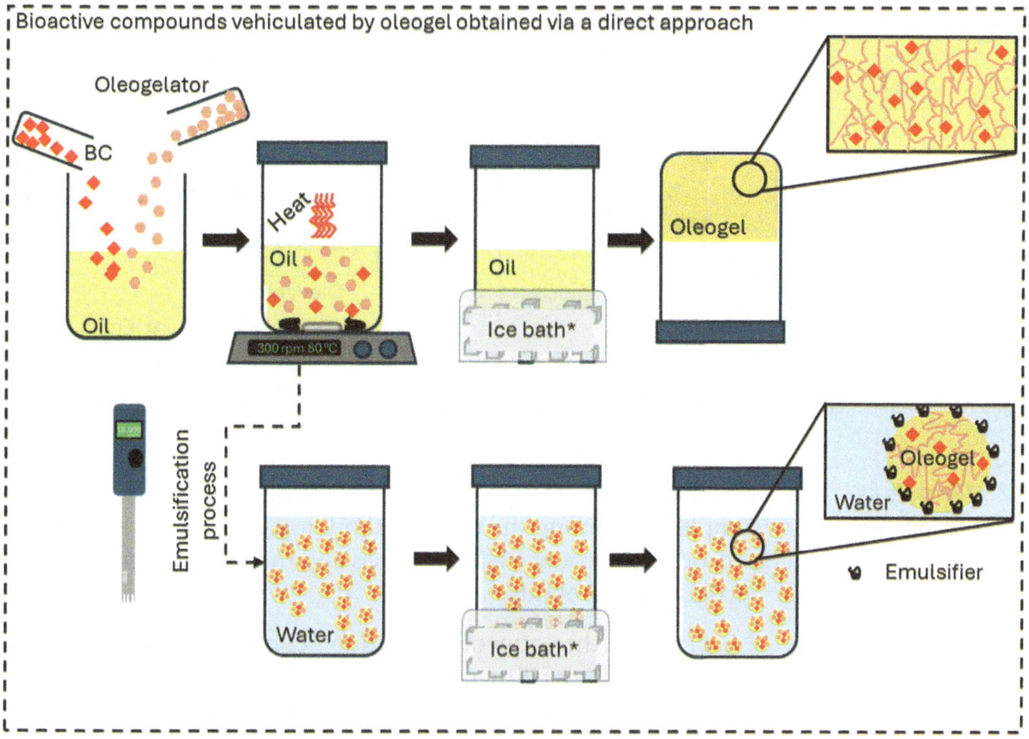

Fig. 2 Schematic steps to use the oleogel obtained by direct approach for delivering bioactive compounds (BC). *The ice bath can be replaced by another cooling system, such as a water bath or an incubation chamber

3. In a separate beaker, prepare an aqueous solution with ultrapure water or buffer and the emulsifier that will stabilize the emulsion (*see* **Note 6**).
4. Heat the aqueous solution for at least 5 min at or above the temperature used to produce the oleogel (*see* **Note 7**).
5. Add the oil + oleogelator mixture loaded with the bioactive compound to the heated aqueous solution.
6. Using a homogenizer (e.g., Ultra-Turrax), mix the two phases (oil and water) at high shear force, at approximately 25,000 rpm, keeping the temperature above the melting temperature of the oleogelator for at least 5 min. If necessary, use a thermostatic circulation bath to maintain the high temperature (*see* **Note 8**).
7. Immediately place the emulsion in an ice bath (~0°C) to form the crystalline network of the oleogel in the droplets dispersed in the emulsion.
8. Store the gelled emulsion in a refrigerator (5°C) for at least 2 h (*see* **Note 9**). These steps are shown at the bottom of Fig. 2.

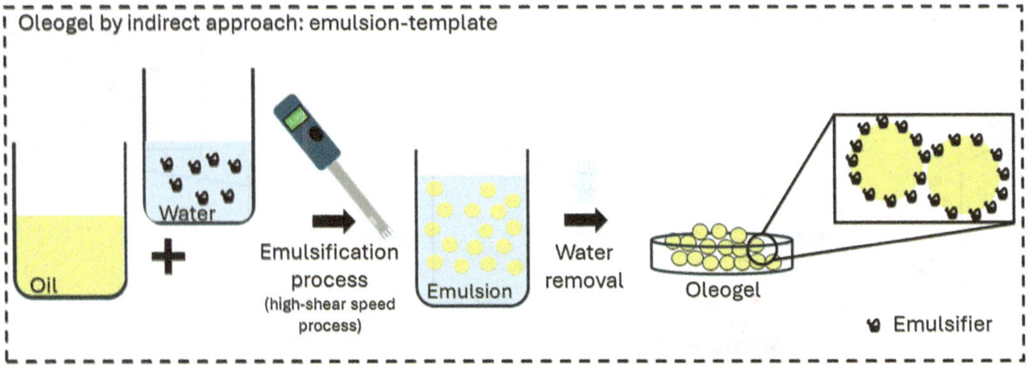

Fig. 3 Schematic steps to obtain the oleogel by emulsion-template

3.2 Oleogels by Indirect Approach

The indirect production of oleogels obtained via emulsion and foam templates is among the main methodologies. This section presents protocols to produce both oleogels and the oleogels enriched with bioactive compounds.

3.2.1 Oleogels by Emulsion-Template

1. Prepare the aqueous solution containing the emulsifier (0.5%–4% g/100 g emulsion) (*see* **Note 10**). If necessary, a buffer solution can be used to solubilize the emulsifier.
2. Homogenize the aqueous solution with the oil phase using a homogenizer (e.g., Ultra-Turrax T18) (*see* **Note 11**).
3. Transfer the stabilized emulsion to a Petri dish, spreading it evenly over the entire surface of the glass with a spatula (*see* **Note 12**).
4. Place the Petri dish with the emulsion in a forced-circulation oven to remove water (*see* **Note 13**).
5. Weigh the Petri dish with the emulsion until it reaches a constant weight.
6. Use a spatula to transfer the dried emulsion into a beaker.
7. Homogenize, manually (~3 min) or using magnetic stirring (1000 rpm, 1 min), the dry emulsion to obtain the oleogel. Figure 3 shows the process steps.

3.2.2 Oleogels by Foam-Template

Hydroxypropylmethylcellulose (HPMC) is commonly found in the literature for the production of this type of oleogel, and therefore, this protocol was based on the production of oleogels using this polymer.

1. Prepare the aqueous solution containing HPMC (*see* **Note 13**).
2. Apply high-shear mixing to the HPMC aqueous solution until foam is formed (*see* **Note 14**).
3. Transfer the aqueous foam to aluminum pans for freezing at −80°C or in liquid nitrogen.

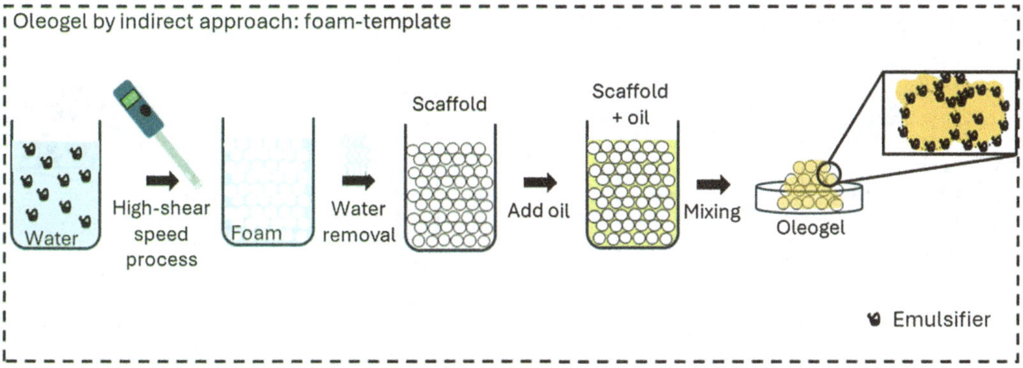

Fig. 4 Schematic steps to obtain the oleogel by foam-template

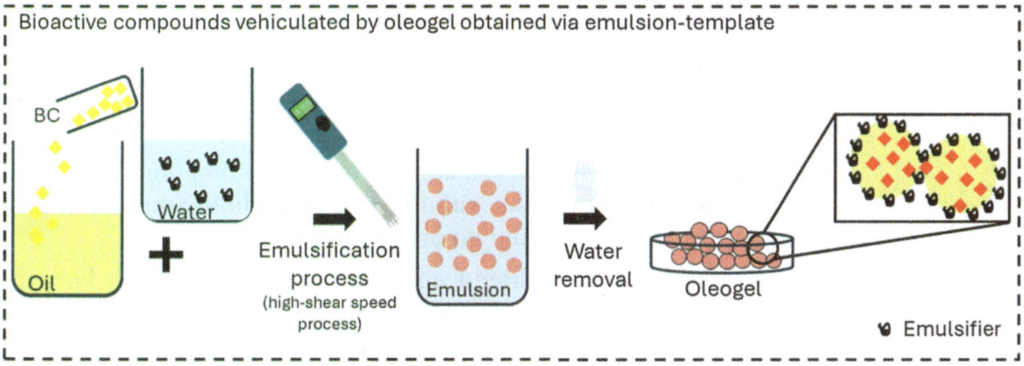

Fig. 5 Schematic steps to use the oleogel obtained by emulsion-template for delivering bioactive compounds (BC)

4. After the foam has frozen, proceed with freeze-drying.
5. Add the dry foam scaffold to a beaker containing oil for absorption.
6. Homogenize the oil-entrapped scaffold manually or use Ultra-Turrax to obtain the oleogel, as shown in Fig. 4.

3.2.3 Incorporation of Bioactive Compounds into Oleogels Obtained by Emulsion-Template

1. Follow Step 1 described in Sect. 3.2.1.
2. Add the lipophilic bioactive compound to the oil that will be used to obtain the oleogel.
3. Keep the oil and bioactive compound under constant stirring until complete dissolution. If necessary, the mixture can be heated to better incorporate the bioactive compound better into the oil.
4. If the oil has been heated, cool it to room temperature before proceeding to the next step.
5. Follow Steps 2–6 described in Sect. 3.2.1. The steps to obtain the oleogel rich in the bioactive compounds are shown in Fig. 5.

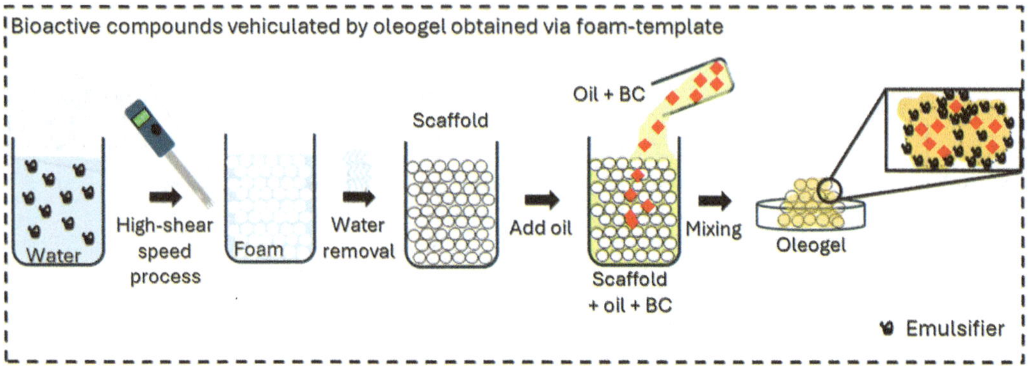

Fig. 6 Schematic steps to use the oleogel obtained by foam-template for delivering bioactive compounds (BC)

3.2.4 Incorporation of Bioactive Compounds in Oleogels Obtained by Foam-Template

1. Follow Steps 1–4 described in Sect. 3.2.2.
2. Add the lipophilic bioactive compound to the oil and keep stirring until it is completely dissolved. If the bioactive compound does not dissolve easily, gentle heating may be needed.
3. Alternatively, you can add the bioactive compound during the mixing of the foam with the oil. If you choose this option, you can skip Step 2 and go directly to Step 4.
4. Add the freeze-dried foam containing the oil and the bioactive compound to the beaker. The scaffold will absorb the oil and bioactive compound at this stage.
5. Follow Step 6 described in Sect. 3.2.2, to obtain the oleogel rich in the bioactive compounds. The process steps are shown in Fig. 6.

4 Notes

1. If necessary to decrease the oil droplet size, high-pressure homogenizer or ultrasound can be adopted.
2. Melting point of some oleogelators used to produce the oleogel. Beeswax: 80°C [8]; carnauba wax: 90°C [9]; glycerol monostearate: 75°C [7]; ethylcellulose: 130°C [10]; γ-oryzanol and β-sitosterol: 80°C [4].
3. The speed of the magnetic stirring will depend on the volume of oil used and should be high enough to better distribute the components in the mixture, but the formation of vortices should be avoided to ensure good homogenization.
4. Oleogel should be stored at temperatures below the melting point of the oleogelator. The choice of temperature will depend on the desired response regarding the behavior of the

oleogel at a given temperature or application. For example, products at room temperature (such as chocolate) are stored at 25°C, and under refrigeration (such as margarine) at 5°C.

5. The use of emulsifiers tends to increase the solubilization of the lipophilic compound in the oil. An example is the use of Span 20 to solubilize β-carotene and curcuminoids in corn oil and MCT [11, 12]. Surfactants can decrease the interfacial tension between the oil, oleogelator, and bioactive compound, preventing the formation of agglomerates. In both studies, the authors used 10% wt of surfactant.
6. It is necessary to establish an emulsifier at the appropriate concentration. For example, to form highly stable emulsions with oleogel and curcumin, Tween 80 and Pluronic F-68 were used at 1.25% and 2.5% w/v, respectively [13].
7. The aqueous solution used to produce the emulsion must have a temperature equal to or higher than that of the oil+ oleogelator mixture to avoid premature crystallization of the oleogelator. Early crystallization may affect the production and stability of the emulsion.
8. Nanoemulsions can be produced using different equipment capable of breaking the emulsion droplets into smaller sizes, such as microfluidizers, high-pressure homogenizers, and ultrasonic devices. The use of ultrasound to produce nanoemulsions has been investigated elsewhere [13]. For example, the ultrasound can be used at medium power (~50%) for approximately 5 min to break up the oil droplets. The important thing is to maintain the emulsion temperature above the melting point of the oleogelator.
9. The storage temperature of the emulsion depends on the desired response to the behavior of the sample at a given temperature.
10. Some emulsifiers used in the literature for stabilizing emulsions to obtain oleogels by indirect approach can be highlighted, such as sodium caseinate, whey protein isolate, xanthan gum, and hydroxypropyl methylcellulose [14–16].
11. The rotation speed and time used to produce the emulsion depend on the emulsifier, oil fraction, and nature of the oil. Therefore, establishing a rotation speed and time for this process is complex. However, studies have used a rotation speed ranging from 11,000 to 16,000 rpm and times ranging from 3 to 5 min [17–20].
12. There is no consensus on the temperature used to remove the aqueous fraction from the emulsion, but temperatures between 45 and 120°C have been used, and the authors have found promising results [17, 21]. Some authors have

also found interesting results using lyophilization to remove water from the emulsion. Cryoprotectants and rapid freezing of samples with liquid nitrogen are recommended for lyophilization.

13. The amount of HPMC used in the studies varies between 1% and 4% wt of water. To dissolve HPMC in water, the solution can be left overnight under stirring.

14. The rotation speed used varies with the processing time; short times require a high value, e.g., 16,500 rpm for 2 min, while lower speeds are used for longer times, e.g., 11,000 rpm for 15 min [2, 22].

References

1. Patel AR, Dewettinck K (2016) Edible oil structuring: an overview and recent updates. Food Funct 7:20–29
2. Oh IK, Lee S (2018) Utilization of foam structured hydroxypropyl methylcellulose for oleogels and their application as a solid fat replacer in muffins. Food Hydrocoll 77:796–802. https://doi.org/10.1016/j.foodhyd.2017.11.022
3. Bascuas S, Salvador A, Hernando I, Quiles A (2020) Designing hydrocolloid-based oleogels with high physical, chemical, and structural stability. Front Sustain Food Syst 4:111. https://doi.org/10.3389/fsufs.2020.00111
4. Martins AJ, Cerqueira MA, Pastrana LM et al (2019) Sterol-based oleogels' characterization envisioning food applications. J Sci Food Agric 99:3318–3325. https://doi.org/10.1002/jsfa.9546
5. Palla C, Valoppi F, (2024). Advances in oleogel development, characterization, and nutritional aspects. Springer International Publishing, Cham
6. Qazi HJ, Ye A, Acevedo-Fani A, Singh H (2024) Delivery of encapsulated bioactive compounds within food matrices to the digestive tract: recent trends and future perspectives. Crit Rev Food Sci Nutr https://doi.org/10.1080/10408398.2024.2353366
7. Barroso NG, Martins AJ, Júnior FDO et al (2024) β-carotene and resveratrol loaded glycerol monostearate-based oleogels: physicochemical characterization at low gelation concentrations. Food Res Int 197:115181. https://doi.org/10.1016/j.foodres.2024.115181
8. Martins AJ, Cerqueira MA, Cunha RL, Vicente AA (2017) Fortified beeswax oleogels: effect of β-carotene on the gel structure and oxidative stability. Food Funct 8:4241–4250. https://doi.org/10.1039/c7fo00953d
9. Lim J, Jeong S, Oh IK, Lee S (2017) Evaluation of soybean oil-carnauba wax oleogels as an alternative to high saturated fat frying media for instant fried noodles. LWT 84:788–794. https://doi.org/10.1016/j.lwt.2017.06.054
10. Laredo T, Barbut S, Marangoni AG (2011) Molecular interactions of polymer oleogelation. Soft Matter 7:2734–2743
11. Cui M, Mao L, Lu Y et al (2019) Effect of monoglyceride content on the solubility and chemical stability of β-carotene in organogels. LWT 106:83–91. https://doi.org/10.1016/j.lwt.2019.02.042
12. Yu H, Shi K, Liu D, Huang Q (2012) Development of a food-grade organogel with high bioaccessibility and loading of curcuminoids. Food Chem 131:48–54. https://doi.org/10.1016/j.foodchem.2011.08.027
13. Palla CA, Aguilera-Garrido A, Carrín ME et al (2022) Preparation of highly stable oleogel-based nanoemulsions for encapsulation and controlled release of curcumin. Food Chem 378. https://doi.org/10.1016/j.foodchem.2022.132132
14. Urbánková L, Sedláček T, Kašpárková V, Bordes R (2021) Formation of oleogels based on emulsions stabilized with cellulose nanocrystals and sodium caseinate. J Colloid Interface Sci 596:245–256. https://doi.org/10.1016/j.jcis.2021.02.104
15. Wang Q, Bobadilla S, Espert M et al (2024) Shortening replacement by hydroxypropyl methylcellulose-based oleogels obtained by different indirect approaches. Texture and sensory properties of baked puff pastry. Food Hydrocoll 153. https://doi.org/10.1016/j.foodhyd.2024.109936
16. Moradabbasi M, Goli SAH, Fayaz G (2022) Effect of biopolymers concentration and

drying methods on physicochemical properties of emulsion-templated oleogel. J Food Sci Technol 59:1994–2003. https://doi.org/10.1007/s13197-021-05214-1

17. Miao W, Zhang Z, Lin Q et al (2025) Preparation of emulsion-template oleogels: tuning properties by controlling initial water content and evaporation method. Food Hydrocoll 158. https://doi.org/10.1016/j.foodhyd.2024.110519
18. Patel AR, Rajarethinem PS, Cludts N et al (2015) Biopolymer-based structuring of liquid oil into soft solids and oleogels using water-continuous emulsions as templates. Langmuir 31:2065–2073. https://doi.org/10.1021/la502829u
19. Tavernier I, Patel AR, Van der Meeren P, Dewettinck K (2017) Emulsion-templated liquid oil structuring with soy protein and soy protein: κ-carrageenan complexes. Food Hydrocoll 65:107–120. https://doi.org/10.1016/j.foodhyd.2016.11.008
20. Zou Y, Xi Y, Pan J et al (2022) Soy oil and SPI based-oleogels structuring with glycerol monolaurate by emulsion-templated approach: preparation, characterization and potential application. Food Chem 397. https://doi.org/10.1016/j.foodchem.2022.133767
21. Santos MAS, Magalhães AER, Okuro PK et al (2024) High internal phase emulsion-template oleogels stabilized by sodium caseinate:quercetin complexes and potential application in pound cakes. J Food Eng 366:111860. https://doi.org/10.1016/j.jfoodeng.2023.111860
22. Wang Q, Espert M, Salvador A, Sanz T (2023) Shortening replacement by emulsion and foam template hydroxypropyl methylcellulose (HPMC)-based oleogels in puff pastry dough. Rheological and texture properties. Curr Res Food Sci 7:100558. https://doi.org/10.1016/j.crfs.2023.100558

Chapter 15

In Situ Polymerization

Gisele Ongmayeb and Ana Silvia Prata

Abstract

In situ polymerization is a robust and scalable technique for the encapsulation of active ingredients, in which polymeric shells are formed directly within the continuous phase of an emulsion. Unlike interfacial polymerization, the reactive monomers are not dissolved in the core material but are instead distributed in the surrounding aqueous medium. This chapter presents a practical protocol for the synthesis of microcapsules using melamine–formaldehyde (MF) resins, with emphasis on pH-responsive and phase-change material (PCM) applications. Despite this type of microcapsules does not meet certain sustainability requirements, there is no proper alternative available. Environmentally conscious adjustments are proposed, including the use of safer pH modifiers and biodegradable emulsifiers, to align with green chemistry principles.

Keywords In situ polymerization, Melamine–formaldehyde, pH-responsive microcapsules, Phase-change materials, Green encapsulation

1 Introduction

In situ polymerization is a chemical encapsulation method that enables the formation of polymeric shells around dispersed core materials without requiring reactive monomers within the core itself. This distinguishes it from interfacial polymerization, where monomers are present in both phases. In situ techniques are particularly suited for systems where the core material is sensitive or incompatible with reactive agents.

The most common class of polymers used in in situ polymerization is represented by aminoplast resins, obtained via polycondensation of urea or melamine (2,4,6-triamino-1,3,5-triazine) with formaldehyde. Formaldehyde as well as the high stability of this shell material are critical environmental factors but there is not yet a more sustainable alternative to match the encapsulation efficiency and mechanical resistance of these classic aminoresin microcapsules [1]. For this reason, the protocol describes the formation of melanin–formaldehyde-based microcapsules.

Ana Silvia Prata and Louise Emy Kurozawa (eds.), *Biopolymers*, Methods and Protocols in Food Science,
https://doi.org/10.1007/978-1-0716-5230-5_15,

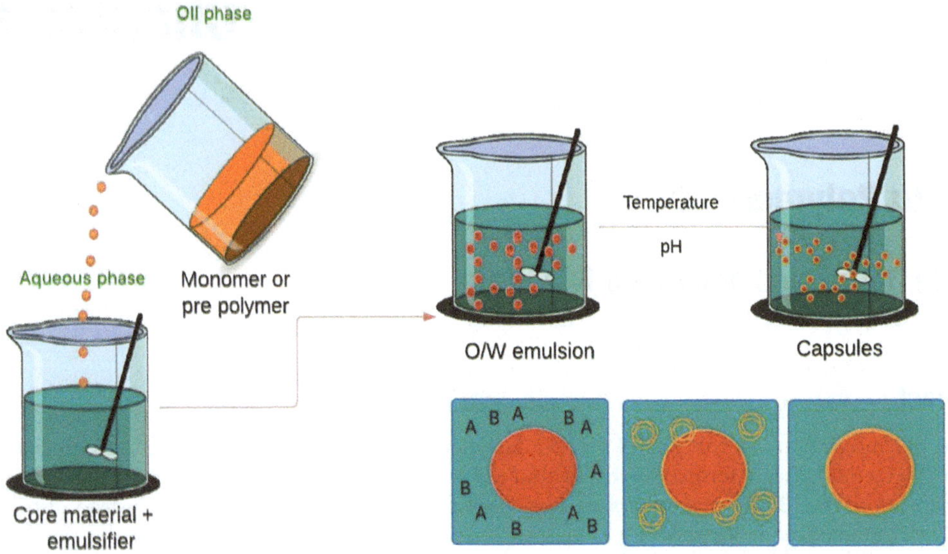

Fig. 1 Formation of microcapsules using in situ polymerization. A and B represent the pre polymer and the crosslinking agent. After adjustment of pH, the polycondensation reaction starts, forming the shell around the oil phase. *Source*: Author

The process typically involves the emulsification of an oil phase (e.g., fragrance or phase-change material [PCM]) into an aqueous medium containing water-soluble monomers, such as melamine and formaldehyde (Fig. 1). Upon acidification, polycondensation is initiated, leading to the formation of crosslinked aminoplast resins that deposit at the oil–water interface, resulting in typical reservoir capsules. These capsules are used in industries ranging from pharmaceuticals and cosmetics to food additives, pesticides, industrial chemicals, and adhesives [2]. The advantage of in situ polymerization is that the responsive molecules may be fixed in the cavity of the polymer due to the long chain or crosslinked network formed inside it, to obtain stimuli responsive particles.

This chapter outlines a reproducible protocol for the synthesis of MF-based microcapsules, including pH-responsive systems and PCM encapsulation. Emphasis is placed on process control, ecological considerations, and formulation adaptability.

2 Materials

Each component was selected for its functional role and ecological compatibility.

- **Prepolymer:** Formaldehyde solution (37%) (*see* **Note 1**).
- **Crosslinking agent:** Melanine.

- **Core oil phase:** Hexadecane (*see* **Note 2**).
- **pH adjustements:** Triethanolamine (TEA) buffer, citric acid solution (2% w/v), and sodium hydroxide solution (0.1 M).
- **Emulsifier:** Styrene maleic anhydride (SMA) copolymer.
- Deionized water.
- Magnetic stirrer with temperature control (*see* **Note 3**).
- pH meter.
- Thermometer or temperature probe.
- Three-necked round-bottom flask.
- Water bath or hot plate.
- Dropping funnel or peristaltic pump.
- Filtration setup (membrane or paper).
- Drying apparatus (room temperature air drying, vacuum oven, or desiccator).

3 Methods

3.1 Prepare the MF Prepolymer

1. In a flask, mix 2.7 g of 37% formaldehyde, 2.0 g of melamine, and 25 mL of distilled water.
2. Adjust the pH to 8–9 by using ~6 drops of TEA.
3. Heat in a water bath at 70°C keeping the stirring (200 rpm) for 60 min.
4. Once transparent, set aside the MF prepolymer.

3.2 Prepare the Core Emulsion

1. Dissolve 0.72 g of SMA in 100 mL of distilled water.
2. Adjust the pH to 4.0 by using ~3 mL of 2% citric acid solution.
3. Add 9.0 g of hexadecane to the emulsifier solution.
4. Stir at 1000 rpm for 30 min at 70°C to form a stable emulsion.

3.3 Polymerization Reaction

1. Maintain emulsion at 70°C, stirring at 350 rpm.
2. Slowly add MF prepolymer over 2 h using a dropping funnel.
3. After addition, adjust the pH to 8–9 with 0.1 M NaOH.
4. Continue the reaction for two more hours at 70°C.

3.4 Recovery and Drying

1. Filter the microcapsule suspension.
2. Wash thoroughly with distilled water.
3. Air-dry at room temperature until a white powder is obtained.

4 Notes

1. Also acrylates, methacrylates, styrenes, or other vinyl monomers, or formaldehyde and urea. More ecological alternatives under study.
2. Phase-change materials: Decane or Rubitherm.
3. Mechanical stirrer or propeller/turbine can also be used.

References

1. Mamusa M, Resta C, Sofroniou C, Baglioni P (2021) Encapsulation of volatile compounds in liquid media: fragrances, flavors, and essential oils in commercial formulations. Adv Colloid Interface Sci 298:102544. https://doi.org/10.1016/j.cis.2021.102544
2. Nguon O, Lagugné-Labarthet F, Brandys FA, Li J, Gillies ER (2017) Microencapsulation by in situ polymerization of amino resins. Polym Rev 58(2):326–375. https://doi.org/10.1080/15583724.2017.1364765

Chapter 16

Interfacial Polycondensation

Gisele Ongmayeb and Ana Silvia Prata

Abstract

Interfacial polycondensation is a versatile technique for the microencapsulation of active compounds, in which complementary monomers react at the interface of immiscible phases to form a polymeric shell around a core material. This method enables the use of a wide range of polymers and core substances, including both hydrophilic and hydrophobic systems. In this chapter, two protocols are presented: one based on an oil-in-water (O/W) emulsion for hydrophobic core encapsulation and another using a water-in-oil (W/O) emulsion to encapsulate hydrophilic cores. The procedures emphasize environmentally conscious formulation strategies, including the use of solvent-free systems and lipid-based excipients, in accordance with green chemistry principles.

Keywords Interfacial polymerization, Solvent-free technology, Melt coatings, Lipid-based excipients, Green encapsulation, Sustainable substrates

1 Introduction

Microencapsulation by interfacial polycondensation involves the formation of a polymeric shell at the boundary between two immiscible phases, typically through the reaction of multifunctional monomers. One monomer is dissolved in the core material, which is then dispersed into a continuous phase usually aqueous containing the complementary reactant. Upon contact at the interface, the monomers undergo rapid polymerization, forming a thin, flexible membrane around the core [1] (Fig. 1).

This technique is rooted in the classical Schotten–Baumann reaction, where an organic-soluble monomer, such as an acid chloride or isocyanate, reacts with a compound containing active hydrogen atoms (e.g., amines or alcohols). The reactivity hierarchy of aqueous monomers typically follows: linear amines > aromatic diamines > primary alcohols > secondary alcohols > phenols [2]. Depending on the monomer pair, various polymer types can be synthesized, including polyurea, polyurethane, polyester, polyamide, polythiourea, polyepoxide, polyacrylamide, and polysiloxane.

Ana Silvia Prata and Louise Emy Kurozawa (eds.), *Biopolymers*, Methods and Protocols in Food Science,
https://doi.org/10.1007/978-1-0716-5230-5_16,

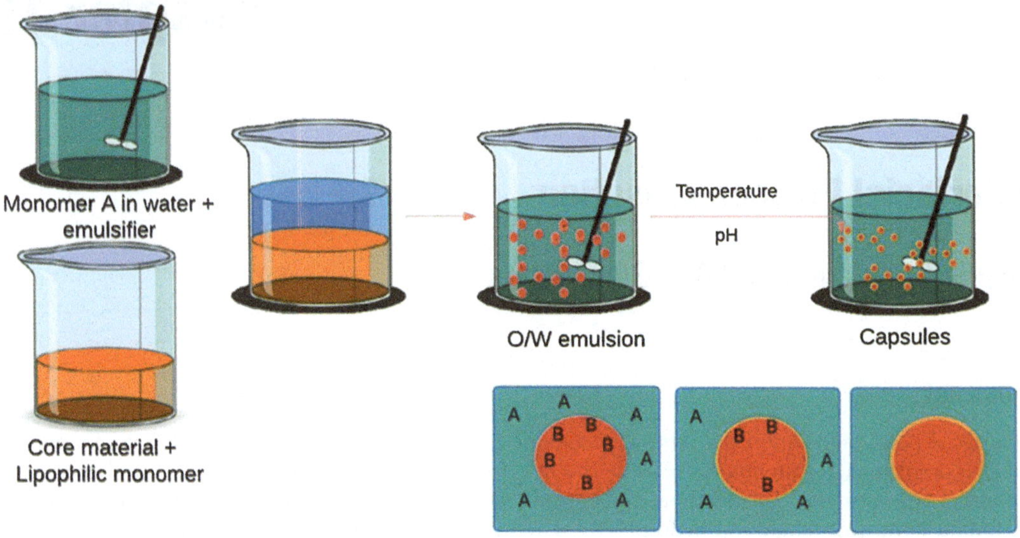

Fig. 1 Schematic representation of interfacial polymerization. Hydrophilic monomer A and lipophilic monomer B. (Source: Author)

The versatility of this method allows for the encapsulation of both hydrophilic and hydrophobic core materials by adjusting the emulsion type—oil-in-water (O/W), water-in-oil (W/O), or oil-in-oil (O/O). However, one limitation is the need to dissolve highly reactive monomers in the core phase, which may compromise sensitive payloads. Additionally, conventional formulations often rely on synthetic solvents and polymers, restricting their use in food or biomedical applications [3, 4].

Commercially, interfacial polycondensation is used to produce microcapsules ranging from 3–6 μm (e.g., for carbonless paper ink) to 20–30 μm in diameter [5]. Its applications span across industries including textiles, cosmetics, pharmaceuticals, agriculture, energy storage, and thermal insulation [6, 7]. In this chapter, we present two reproducible protocols for encapsulating hydrophobic and hydrophilic cores, with an emphasis on greener formulation strategies and scalable process design.

2 Materials

2.1 Equipment

- Magnetic stirrer with temperature control.
- Thermometer or temperature probe.
- Overhead mechanical stirrer.
- Droplet addition system (e.g., syringe pump or burette).

- Vacuum drying apparatus.
- Centrifuge.
- Beaker 50 mL, 100 mL.
- Cold bath.

2.2 Materials for Hydrophobic Core Encapsulation

Ingredients for 100 mL of emulsion:

- 2 g of Tween 80 emulsifier (*see* **Note 1**).
- 0.5 g of a hydrophilic monomer B (*see* **Note 2**).
- 5 mL of hydrophobic core active ingredient (*see* **Note 3**).
- 20 mL of an organic solvent (*see* **Note 4**).
- 1 g of diacid chloride (monomer A) (*see* **Note 5**).
- 75 mL of deionized water.

2.3 Materials for Hydrophilic Core Encapsulation

Ingredients for 100 mL of emulsion:

- 2 g of lecithin emulsifier (*see* **Note 6**).
- 0.5 g of a hydrophilic monomer B (*see* **Note 7**).
- 10 mL of hydrophilic core active ingredient (*see* **Note 8**).
- 80 mL of an organic solvent (*see* **Note 9**).
- 1 g of diacid chloride (monomer A).

3 Methods

3.1 Hydrophobic Core Encapsulation

1. In a beaker, prepare the aqueous phase by mixing 75 mL of deionized water with 2 g of Tween 80.
2. Stir until fully dissolved (≈5 min).
3. Prepare the organic phase by dissolving 1 g of diacid chloride in 20 mL of ethyl lactate.
4. Then, add to this phase 5 mL of the hydrophobic active core and mix gently.
5. Create the emulsion by pouring slowly the organic phase into the aqueous phase under high-speed stirring (≈10,000 rpm).
6. Emulsify for 5–10 min until droplets are uniform.
7. Initiate polymerization by adding 0.5 g of ethylenediamine dropwise to the emulsion.
8. Maintain stirring at 30°C for 2–4 h.
9. Recover and purify the particles

- Cool to room temperature.
- Centrifuge at 3000 rpm for 10 min.
- Wash the pellet 3× with deionized water.

10. Dry and store

- Dry under vacuum at 35°C for 12–24 h.
- Store in airtight containers, preferably refrigerated.

3.2 Hydrophilic Core Encapsulation

1. Prepare the aqueous core by dissolving 0.5 g of ethylenediamine in 10 mL of deionized water.
2. Add the hydrophilic active ingredient and mix gently.
3. Prepare the organic phase by dissolving 1 g of diacid chloride in 80 mL of limonene.
4. Then, add 2 g of lecithin and stir until fully dispersed.
5. Create the emulsion by slowly adding the aqueous core into the organic phase under moderate stirring (≈5000 rpm).
6. Emulsify for 10 min until droplets are stable.
7. Initiate polymerization

- Maintain stirring at 30–35°C for 3 h.
- Monitor viscosity and droplet stability.

8. Recover and purify

- Cool to room temperature.
- Centrifuge at 3000 rpm for 10 min.
- Wash the pellet with ethanol and water to remove residuals.

9. Dry and store

- Dry under vacuum at 35°C.
- Store in sealed containers, protected from light and humidity.

4 Notes

1. The protocol describes the amount of a nonionic, biodegradable surfactant. Lecithin can also be employed. Traditional surfactants, such as Amisoft GCS 11 (anionic emulsifier) and poly(vinyl alcohol) (PVA, 1500 MW), are used in other concentrations to that described in this protocol.
2. Hydrophilic monomer is used to react at the emulsion interface. Normally, a multifunctional amine or a combination of

amines, such as ethylenediamine, 1,6-hexamethylenediamine, or triethylenetetramine.

3. Natural extract preferred, such as an essential oil. If aiming for the use of a solid core material, it must be dissolved in a water-immiscible solvent before the monomers or prepolymers are added.
4. Organic solvents must be biodegradable with low toxicity. Limonene can also be employed.
5. A hydrophobic monomer must be used, such as a multifunctional acid chloride (e.g., sebacoyl chloride, soluble in green solvent). Alternatively, isocyanate such as 2,4-toluene diisocyanate or polyisocyanate monomer can be used isolated or in combination with acid chloride.
6. Lecithin, lipid based, stabilizes W/O emulsions. Span 80 can also be used.
7. Ethylenediamine, soluble in aqueous core.
8. Aqueous extract, water soluble, sensitive.
9. Limonene or propylene carbonate, biodegradable, low volatility.

References

1. Markus A, Linder C (2006) Chapter 2 - Advances in the technology for controlled-release pesticide formulations. In: Benita S (Ed.) Microencapsulation Methods and Industrial Applications. 2nd edn. CRC Press, Boca Raton, 54–78. https://doi.org/10.1201/9781420027990
2. Perignon N et al. (2014) Interfacial polymerization for microencapsulation: a review of mechanisms and applications. J Microencapsul 31(6):537–553
3. Clark JH, Tavener SJ (2007) Alternative solvents: shades of green. Green Chem 9(5):438–448
4. Kerton FM, Marriot R (2013) Alternative solvents for green chemistry. Royal Society of Chemistry. 2nd ed. 350p. https://doi.org/10.1039/9781849736824
5. Arshady R (1989) Microparticles prepared by interfacial polymerization techniques. J Control Release 10(1):1–35
6. Jyothi NVN, Prasanna PM, Sakarkar SN, Suresh B (2010) Microencapsulation techniques and its applications in the pharmaceutical field: a review. J Pharm Res 3(1):26–32
7. Wang W, Wu Y, Li J, Wang Y (2011) Microencapsulation of essential oils with biodegradable polymers for food and cosmetic applications. Food Res Int 44(9):2782–2790

Check for updates

Chapter 17

General Protocol for Particles Characterization

Ana Silvia Prata

Abstract

This protocol aims to concentrate the main characterization techniques of the particles described in the Book of Protocols, serving as a guide for encapsulation procedures. A general characterization involves determining moisture content, the behavior of solubility in a media, and the density of particles and bulk of particles. Morphological analyses for moist and dry samples, as well as core and polymer distributions, are described. The main size particle techniques for large or small particles include light scattering, image treatment, and sieving. Finally, encapsulation efficiencies are described in two parts: extraction of the core material and quantification techniques. With this compilation, most of the particles used for the encapsulation of bioactive compounds can be roughly characterized.

Keywords Size, Morphology, Encapsulation efficiency, Wettability, Density

1 Introduction

Different encapsulation techniques developed throughout scientific history after carbonless copy paper, have resulted in particles with different active-compound distributions, sizes, solubility, encapsulation efficiencies, and release characteristics.

The most relevant information about a particle relates to these characteristics, and in general, they have been grouped together in this chapter.

These techniques have advantages and limitations, some of which are compiled in Table 1, serving as guidance for solving specific problems or for developing new encapsulation-based systems.

The size of particles influences the sensorial perception in the product, regardless of the application. Smaller sizes lead to a smooth mouthfeel or skinfeel, better spreadability, and may change the visual aspect of the product. Micro-sized pigments (e.g., TiO_2 100–200 nm) used in sunscreen creams, for example, are opaque or white, while the same nano-sized pigments (<100 nm) are transparent and maintain the UV protection. The size influences the stability of the product, such as those based on emulsions, and it is directly

Ana Silvia Prata and Louise Emy Kurozawa (eds.), *Biopolymers*, Methods and Protocols in Food Science,
https://doi.org/10.1007/978-1-0716-5230-5_17,

Table 1
General aspects of encapsulation techniques

Method	Advantages	Limitations	Particles (μm)	Cost	Encapsulation efficiency (%)
Spray drying	Scalable and continuous; suitable for heat-stable materials; flexible; equipment availability	Depends on glass transition of wall materials, and may cause core material degradationor displacement of small molecules to the surface	Dry form; Most matricial and hydrosoluble, 1–60	Low	30–80
Hydrosoluble coating (air suspension)	Uniform coating; good for sensitive materials; controlled release possible; scalable	Limited to solid cores; coating uniformity depends on process control; evaporation depends on the air conditions.	Dry form; For coating, 150–5000	Medium–High	50–90
Hot melting (air suspension)	Solvent-free; suitable for lipid-based coatings; good protection against oxidation	Thermal degradation risk; limited to heat-stable actives	Dry form; For coating, 150–5000	Medium	50–90
Thermoplastic Extrusion	Continuous process; good encapsulation of flavors and vitamins; solvent-free; scalable	Requires heat-stable compounds; limited flexibility in carrier materials; small final size depends on milling	Dry form; 500–5000	Medium	40–95
Ionic Gelation	Mild conditions; suitable for high MW bioactives and probiotics; biodegradable polymers	Poor mechanical strength; sensitive to pH and ionic environment. Porous structure	Hydrogel; Matricial; 400–3000	Low–Medium	60–95
Complex Coacervation	High loading capacity; good protection; suitable for hydrophobic compounds	Sensitive to pH and ionic strength; expensive polymers (e.g., gelatin, gum arabic); batch production; crosslinking control	Hydrogel; Matricial or reservoir;10–80	Medium–High	70–98

Table 1
(continued)

Method	Advantages	Limitations	Particles (μm)	Cost	Encapsulation efficiency (%)
In situ polymerization	Strong and stable capsules; good for volatile and reactive compounds	Toxic monomers or reagents may be required; requires purification	Dry; reservoir; 0.1–10	Medium–High	60–95
Interfacial polymerization	Thin, robust capsule walls; control over permeability	Use of organic solvents; potential toxicity; complex process control	Dry; reservoir; 0.1–10	High	70–95
Solvent evaporation	Simple; suitable for hydrophobic compounds; allows control of particle size; mild conditions for actives	Use of organic solvents; possible residual solvent; limited to oil-in-water or water-in-oil systems Difficult to scale up	Dry; 0.1–100	Medium	60–95
Emulsion-based encapsulation	Simple and versatile; tunable droplet size; suitable for lipophilic compounds	Stability issues (coalescence, creaming); limited control over release	Wet; 0.1–50	Low–Medium	50–90
Spray chilling (or cooling)	Solvent-free; protects sensitive actives from oxidation and light; quick solidification	Not suitable for hydrophilic actives; limited thermal stability range	Solid; 80–1000	Low–Medium	50–90
Structured lipid particles (e.g., SLN, NLC)	Excellent protection; controlled release; suitable for lipophilic actives	Complex formulation; possible polymorphic transitions; high equipment cost	Solid; 0.05–100	Medium–High	70–95

related to the release rate of the core material and its protection. Greater surface area facilitates oxidation or degradation reactions, the enzymatic action, or dissolution occurring in the initial stages of digestion. The size is tuned to improve the functionality of the product, giving deeper penetration in the skin, passing unnoticed by the body's defense cells, or to provide a delayed release of probiotics or minerals, for example.

The determination of the load a particle is the most difficult analysis to perform due to uncertainties during the core material extraction procedure, which can have limited partition in the solvent, to keep "attached" to the wall material or simply to be lost, since it is in very low quantity. Moreover, quantification techniques have limitations related to standards adopted, complexity of the core material, and sensitivity of the technique. But this characterization is the "core" of encapsulation, influencing the efficacy of the formulation. However, some encapsulation techniques are known to result in higher Encapsulation Efficiencies, as illustrated in Fig. 1.

The core distribution is another important information because it confirms the location of a probiotic strain, an enzyme for reactions or the active compound for further release.

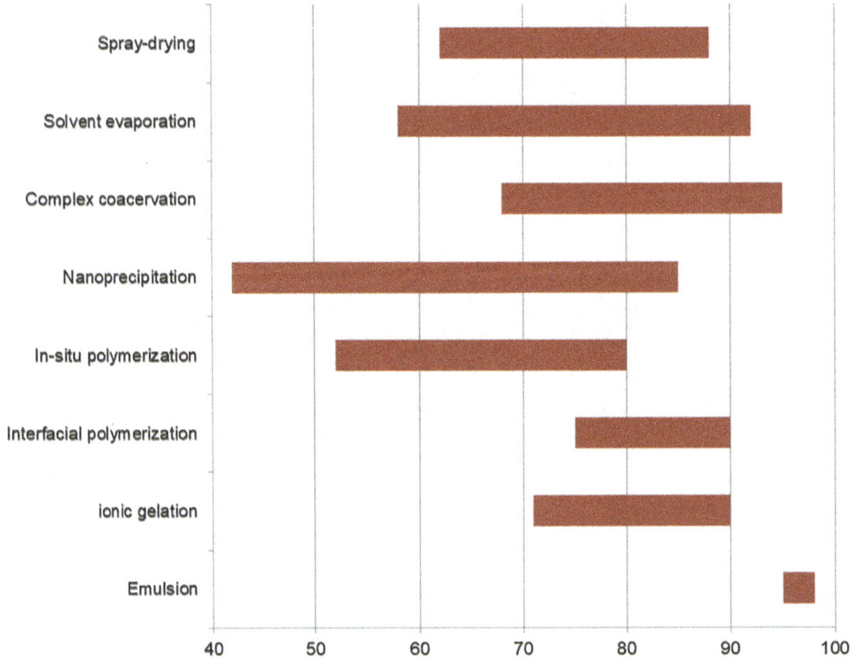

Fig. 1 Efficiency encapsulation of several techniques

2 Materials

2.1 Basic Analysis

- Distilled water.
- Non-penetrating solvent (e.g., kerosene or mineral oil depending on sample compatibility).
- Drying oven (capable of 105°C).
- Vacuum oven (50°C, 93 mmHg)—for heat-sensitive samples.
- Analytical balance (precision ±0.0001 g).
- Heat-resistant weighing dishes or aluminum pans.
- Desiccator (to cool samples to room temperature before weighing).
- Stirring plate with speed control (600 rpm).
- Magnetic stir bars.
- Centrifuge capable of 3000 × *g*.
- Centrifuge tubes (≥50 mL).
- Pipettes.
- 100 mL graduated, tared cylinder.

2.2 Morphology

- Microscope slides (25.4 × 76.2 mm).
- Pasteur pipettes.
- Aluminum stubs.
- Double-sided sticky tape.
- Gold sputter-coater.
- 2.5% glutaraldehyde in 0.1 M phosphate buffer (pH 7.4).
- Ethanol solutions for dehydration.
- Nile red solution 0.1 mg/mL in acetone or in dimethyl sulfoxide (DMSO).
- FITC (Fluorescein isothiocyanate) stock solution (1 mg/mL in DMSO or 0.2 mg/mL ethanol). FITC is used for proteins. For starch, Nile blue A (100 μL, 0.1%, dissolved in distilled water), Calcofluor White (0.1 g/L in ethanol) for carbohydrates.
- Optical microscope with camera.
- Software for processing the images.
- Scanning Electronic Microscope.
- Confocal microscope (Nile Red → excitation ~543–552 nm and emission at 610–710 nm, FITC → excitation 488 nm and emission at 498–547 nm, RBITC → excitation ~543 nm, **Calcofluor White**: Excitation at 405 nm and emission at 410–490 nm).

2.3 Size of Microparticles

- Microscope slides (25.4 × 76.2 mm);
- Pasteur pipettes.
- Optical microscope with camera.
- Software for processing and analyzing the size of the microparticles (ImageJ).
- Dispersion medium (water or other solvents, depending on the sample nature).
- Sample preparation container (e.g., beakers, flasks).
- Deionized water (or another appropriate dispersion medium).
- Appropriate dispersing agents (optional, if needed to prevent aggregation of particles).
- Mastersizer instrument (e.g., Malvern Instruments, or equivalent).
- Wet sample (dispersion of particles in a liquid, typically water or a suitable solvent).
- Ultrasonic bath or probe (optional, for sample dispersion if required).
- Software for processing and analyzing the size of the microparticles.
- Set of sieves with selected mesh sizes.
- Sieve shaker (vibratory or mechanical).
- Precision balance (0.001 g or better).

2.4 Encapsulation Efficiency

- Organic solvents (depending on method):
  - o Petroleum ether.
  - o Hexane.
  - o Ethanol (95%).
  - o Methanol.
  - o Acetone.
  - o Chloroform.
  - o Other water-immiscible solvents.
- Water (for water-soluble particle solubilization).
- Protease solution (0.5 mg/mL; adjustable concentration).
- Hydrochloric acid (2.0 mol/L).
- Sodium citrate solution (0.2 M).
- Internal standard (for GC analysis).
- Filter paper (Whatman No. 1 or similar).
- Syringe filters (0.22 μm) or cotton plug (optional for filtering using syringe).

- Volumetric flasks (10, 25, 50, 100 mL).
- Separation funnels (250–500 mL).
- Petri dishes (tared, dried).
- Porcelain mortar and pestle.
- Antifoam (silicone oil).
- Analytical balance (±0.001 g).
- Incubator or water bath (37°C).
- Vortex mixer or gentle shaker.
- Heating plate (if boiling lipid-based particles).
- Clevenger hydrodistillation apparatus.
- Micropipettes.
- Centrifuge tubes.
- Gas chromatography vials (amber preferred).
- Rotary evaporator with vacuum controller.
- Water bath.
- Standard active compound (essential oil, vitamin, etc.)
- Cuvettes (quartz or glass, depending on wavelength).
- UV–Vis spectrophotometer (190–500 nm range).
- External standards (e.g., limonene) at known concentrations.
- Gas Chromatograph.

3 Methods

3.1 Basic Characterization

Moisture content: Dry of a mass of sample (about 2000 g) in an oven until constant weight at 105°C or under vacuum (93 mmHg, 50°C) to sensitive compounds. Moisture content (%) is the ratio of weight loss and total weight.

$$\textit{Moisture content}\ (\%) = \frac{(\textit{wet particle mass} - \textit{dry particle mass})}{\textit{wet particle mass}} \times 100$$

Water solubility: Add 10,000 g of sample to cold (room temperature) or boiling distilled water (25 mL). Stir at 600 rpm for 10 min. Centrifuge at 3000 × *g* for 10 min. Determine the moisture content of the supernatant. The water solubility is the ratio of the solid in the upper supernatant and the weight of the dry sample used.

Wettability: Sprinkle 10,000 g of powder over the surface of 100 mL of distilled water at 20°C without agitation. Record the time taken for the powder particles to sediment, sink, be submerged, and disappear from the water's surface.

Bulk density (ρd) and tapped density (ρb)

Load a 100 mL-tared graduated cylinder with powder. Read the volume occupied directly from the cylinder (V1). Weigh the mass of particles (P1). The mass/volume relationship is bulk density (ρd). Tap the cylinder repeatedly by lifting and dropping it under its own weight until a negligible difference in volume between successive measurements is observed. Read the new volume (V2). This new mass/volume relationship is the bulk tapped density (ρt).

Add a known volume of a solvent to the graduated cylinder. This solvent should not be able to penetrate the particles. If needed, mix the solvent quickly to fill all gaps between the particles. Measure the total volume (V3), which represents the solvent's displacement volume caused by the presence of particles. Calculate the real particle volume (V3) by V3 = V2−V1. Calculate the real density of particles ρr = P1/V3.

$$\rho b = \frac{particle\ mass}{occupied\ volume}\quad \rho t = \frac{particle\ mass}{tapped\ volume}\quad \rho r = \frac{particle\ mass}{displaced\ volume}$$

3.2 Morphology

3.2.1 Optical Microscopy

1. Place an aliquot of particle suspension on a glass slide by using a Pasteur pipette. If particles are wet, mix them before collecting the sample (*see* **Notes 1 and 2**).
2. Adjust the microscope to observe the morphology of the microparticles under 10 × and 40 × objectives using an optical microscope coupled to a digital camera (*see* **Note 3**).

3.2.2 Scanning Electron Microscopy (SEM) Analysis

1. For tissues or biological materials:
   a. Immerse the biological sample in the fixative medium (e.g., 2.5% glutaraldehyde in 0.1 M phosphate buffer [pH 7.4]) for 2–4 h at 4 °C.
   b. Rinse 3 × with phosphate buffer (10 min each).
   c. Post-fix in 1% osmium tetroxide for 1 h at room temperature (enhances contrast and stabilizes lipids).
   d. Rinse 3 × with distilled water.
   e. Dehydrate the sample in a gradient ethanol series 30, 50, 70, 90, and 100% ethanol.
   f. Immerse in each for 10–15 min.
   g. Repeat 100% ethanol step 2 × to ensure full dehydration.
2. For most samples
   a. Dry the sample under air drying or freeze drying (*see* **Notes 4 and 5**).

3. Place the sample on the adhesive tape over aluminum stubs (*see* **Note 6**).
4. Use a sputter coater to deposit a thin conductive layer (~10 nm) of gold, gold-palladium, or platinum.
5. Coat uniformly to avoid image artifacts.
6. Place the sample stub in the SEM chamber.
7. Set appropriate parameters:
8. Electron beam voltage: most common 5 and 10 kV (1–20 kV, depending on sample).
9. Magnifications: × 1000, × 3000, and × 4000.
10. Choose suitable detectors: SE (secondary electron) for surface morphology or BSE (backscattered electron) for composition contrast.

3.2.3 Confocal Laser Scanning Microscopy (CLSM) Analysis

Preparation of Dye Solutions

Blue coloration for carbohydrates: Calcofluor White (1 mg/mL)

- Using a pipette, add 1000 μL of Calcofluor White to a 10 mL volumetric flask.
- Complete the volume to 10 mL with ethanol.
- Store at 4°C, protected from light.

Green coloration for proteins: FITC (0.2 mg/mL)

- Weigh 10 mg of FITC into a 10 mL beaker.
- Add approximately 7 mL of ethanol and transfer to a 50 mL volumetric flask.
- Repeat this procedure at least three more times to ensure complete transfer of the dye.
- Complete the volume with ethanol.
- Store at 4°C, protected from light.

Red coloration for oil: Nile Red (0.1 mg/mL)

- Weigh 10 mg of Nile Red into a 10 mL beaker.
- Add approximately 7 mL of acetone and transfer to a 100 mL volumetric flask.
- Repeat this procedure at least three more times to ensure complete transfer of the dye.
- Complete the volume with acetone.
- Store at 4°C, protected from light.

Fluorescence labeling of the oil phase by Nile red

- Dissolve the Nile Red in the oil phase to at least 1 mg/mL of oil.

Fluorescence labeling of protein

- Dissolve the FITC C or RBITC in DMSO (1 mg/mL).
- Add 100 μL of the dye solution to the polymer solution (2.5% w/v, pH 8–11) and stir for 1 h at 40°C.
- Add 50 μL thanolamine to stop the reaction.

1. Produce the particles with the colored oil phase.
2. Wash the particles with ethanol to remove the excess or free dye
   - For emulsions:
   - Add 1 mL of fresh emulsion and 10 μL of each dye to a 1.5 mL microtube.
   - Homogenize using a vortex for 30 s.
3. Carefully place the stained sample on a glass slide.
4. Add a small drop of mounting medium to preserve fluorescence.
5. Place a coverslip on top, avoiding air bubbles.
6. Seal the edges with clear nail polish or mounting sealant (optional).
7. Lipid Visualization
   - Adopt one fluorescence channel in the single red fluorescence mode at 514 nm excitation and with the filter block A1.
   - The final pictures are composed from the mixer A (red fluorescence) and B (transmitted light) to visualize oil phase and capsule wall in the same image.
8. Capsule wall structure
   - Adjust the laser in the green/red fluorescence mode (two excitation wavelengths at 488 and 514 nm).
   - Use appropriate filter settings, such as the emission filter blocks VHS/A1 and A2.
   - Green and red fluorescence images are acquired from two separate channels, and a third picture from the transmitted light detector is optional.

3.3 Particle Size Distribuiton (PSD)

PSD is expressed in the form of polydispersity (PdI), also called the coefficient of variation, calculated as the quotient of standard deviation σ and z-average (Zd):

$$\mathrm{PdI}(\%) = \sigma / \mathrm{Zd} * 100$$

3.3.1 Size of Particles Smaller than 800 um

1. In a static light scattering instrument, switch on the dispersion unit(s) and follow the manufacturer's protocol to realize the measurements.
   a. Dry accessory is an alternative to the wet measurements, suitable for dry particles. In this case, fill the sample compartment.
2. Ensure that particles will not settle or be dissolved with the chosen media.
3. Switch on the computer and start the software.
4. The parameters inserted in the program are the refraction index of the media, type of model particles and theory adopted for calculate the size of particle. The theory for size calculation assumes spherical particles.
   a. Large and opaque particles are usually analyzed using Fraunhofer diffraction theory. Mie theory also considers other optical phenomena in addition to diffraction.
   b. D_{50} defines the point where 50% of the particles are smaller and 50% bigger than that certain diameter. The beginning and end of the distribution are commonly defined by D_{10} and D_{90}.
   c. Percentages of the particle sizes in the sample are typically given in volume, but the results can also be chosen for number-based distributions.
5. Dilute the particles in ultrapure water (1:100 w/w) to eliminate multiple spreading and viscosity effects caused by the concentrated composition. Non-ionic surfactant can be added (1.5% wt.) to help the dispersion. Depending on the particle solubility, ethanol or isopropanol is used.
6. Place the sample in the appropriate cuvettes.
7. Insert the cuvette in the instrument.

3.3.2 Size of Large Particles

Optical microscopy

1. Take several pictures from different positions in the glass slides through the optical microscope (see how to prepare and capture in 3.2.1)
2. Launch ImageJ software.
3. Go to File → Open and select the image file to open the image to be treated.
4. Convert the Image to Grayscale (if necessary)
   a. Go to Image → Type > RGB Stack (to separate the color channels) and then convert each to grayscale by selecting Image → Type → 8-bit.

5. Go to Image → Adjust → Threshold, to adjust the sliders (either Lower or Upper) until the particles are clearly separated from the background. You should see the particles appear in a binary form (black particles on a white background).
6. Click Apply to set the threshold. This will convert the image into a binary format, making it easier to analyze particles.
7. Set the Measurement Scale
   a. Go to Analyze → Set Scale
   b. Draw a line between two points of known distance such as a ruler on the photograph to define the pixel corresponding to the length of the line (*see* **Note 7**).
   c. Type the known distance and units of measure (e.g., micrometers or millimeters) in the appropriate boxes and click OK.
   d. Measurements will now be shown using these settings.
8. Go to Analyze → Analyze Particles.
   a. Type the upper and lower limits for the particle sizeparticle size (e.g., 0-Infinity for all sizes, or specify a size range if you only want to measure particles within a certain size range).
   b. Toggle "show outlines"
   c. Check the "Display Results" for outlines of the particles on the image (useful for visual confirmation).
   d. Click OK.
   e. Select Add to Manager if you want to add the measurements to the Results table.
   f. The data window contains measurements for each particle, including Area, Perimeter, Circularity, and Shape descriptors.
9. Go to File > Save As and choose the desired format (e.g., CSV, Excel) for further analysis.

Sieving

1. Choose a sequence of mesh sizes of sieves that can represent the sample size distribution (*see* **Note 8**).
2. Mount the sequence of sieves in the appropriate support to form a column in progressively smaller mesh size from up to bottom (*see* **Note 9**).
3. Place the sieve column in a sieve shaker.
4. Insert a pre- weighed mass of particles (more than 100 g).

5. Turn on the vibration at a specified frequency for a set duration (typically 10 min).
6. Carefully pour each fraction into a capsule for weighing. Special attention should be given to removing any remaining grains from the sieve, using a brush suited to the fineness of the mesh. Brushes should never be used to press down on the sieve, and grains trapped in the mesh should not be forcibly removed.
7. Weigh each sieve fraction by using a precision balance.
8. Record the measurements on paper and then enter into a spreadsheet for statistical analysis.

3.4 Encapsulation Efficiency Analysis

Definitions

Encapsulation efficiency (EE): The amount of core material encapsulated inside the particles compared to the initial core material (IO) used to produce the particles. Alternatively, it can be calculated by using the total oil (TO) or the total oil–surface oil.

Surface oil (SO) is related to the unencapsulated oil at the surface of the particles, and it is measured mostly through solvent extraction. The contact time as well as the detection methodology used will determine this parameter.

$$EE(\%) = \frac{mass\ of\ encapsulatd\ oil}{mass\ of\ oil\ in\ process}$$

EE = TO/IO or TO-SO/IO

EE = load of particles/ theoretical load

Both loads are on a dry-weight basis.

Load: The amount of core material encapsulated inside the particles (g of active compound)/g of particles

$$Load = \frac{core\ material\ mass}{particle\ mass}$$

The following sections are divided into extraction and quantification

3.5 Liquid–Liquid Extraction After Breaking The Particle Wall

For water-soluble particles, weigh an appropriate amount of the microparticle sample (e.g., 1000 g or 5000 g) and transfer it into a beaker.

Add it to a sufficient volume of water (e.g., 20 mL) and ensure the solubilization of the wall material.

For water-insoluble particles, evaluate which condition can break the wall.

If the particle is a polymeric complex without chemical crosslinking: Weigh 0.5 g of microcapsule and add it to 5.0 mL of HCl solution

(2.0 mol/L) (*see* **Note 10**). Incubate this solution at 37°C for 10 min (if not dissolved, then shake it gently).

If the particle contains protein: Weigh 0.5 g of microcapsule and add it to 5.0 mL of a protease solution (0.5 mg/mL (increase concentration as needed) in water). Incubate this solution at the optimum temperature of the enzyme for 10 min (if not dissolved, then shake it gently) (*see* **Note 11**).

If the particle is a hydrogel crosslinked with calcium ions: Weigh 0.5 g of microcapsule and add it to 5.0 mL of a sodium citrate solution (0.2 M). Incubate this solution at 37°C for 10 min (if not dissolved, then shake it gently) (*see* **Note 12**).

If the particle is based on lipid: Boil the mixture water-particle at the melting point of the lipid while stirring continuously for 45 min. After boiling, allow the mixture to cool to room temperature. Alternatively, lipid particles can be dissolved in an organic solvent (methanol, acetone). In this case, the organic solvent + active compound is directly quantified by 3.9.2 or 3.9.3.

1. To any of the solutions of particles, add 20.0 mL of petroleum ether or other water-immiscible solvent in a Separation funnel together with the aqueous solution (*see* **Note 13**).
2. Gently shake the funnel with the mixture for about 10 min. Open the funnel sometimes to release the vapor (*see* **Note 14**).
3. Allow the mixture to settle and the phases to separate clearly. This may take 10–15 min or longer, depending on the solvent system used.
4. Carefully separate the phases and recover the desired one (separation by draining).
5. Transfer this phase to a volumetric clean flask.
6. Repeat the extraction process twice and combine the phases into the same flask.
7. If microparticles are still present in the extract, filter the liquid through filter paper or a syringe filter (e.g., 0.22 µm filter) to ensure the final extract is free from particulate matter.
8. See the procedure to quantify by gravimetry, chromatography or spectrophotometry.

3.6 Solvent Extraction by Crushing the Particle

1. Add 20,000 g of wet microparticles in dried tared Petri dishes in triplicate.
2. Remove the moisture from the microparticles by placing the plates in a vacuum oven at 50°C for 24 h.
3. Weigh the resulting dried microparticle sample in a Petri dish.
4. Add the solvent able to solubilize the core material (e.g., 95% ethanol) to the plates in order to slightly remove the solids that are present on them (*see* **Note 15**).

5. Transfer the content to a small porcelain mortar for grinding.
6. Drop small quantities of the solvent, and ground vigorously the particle with a pistille.
7. Extract the supernatant by using a syringe with a 0.22 μm filter attached to its fine end (or a small cotton piece in the tip) ensuring that no particles are dragged.
8. Collect each extract in a volumetric flask.
9. Repeat this process until a volume of 25 to 30 mL of the supernatant is obtained.
10. Finally, complete the volume of the volumetric flask with fresh solvent.
11. Alternatively, at the end of sequential maceration, the total volume and the crushed particles are transferred to a flask with tip, and stirred at 300 rpm for 30 min. The solution is then filtered through a cellulose filter. The retained material is washed with 5 mL of solvent three times.
12. See the procedure to quantify by gravimetry, chromatography or spectrophotometry.

3.7 Hydrodestillation for Oil Recovery

1. Dissolve 10 g (± 0.02 g) of the water-soluble particles in 250 mL of deionized water by using a 500-mL round-bottomed flask.
2. Manually shake the flask for approximately 1 min.
3. Add 0.5 mL of antifoam solution (silicon oil).
4. Fit the Clevenger apparatus to the top of flask. The Clevenger apparatus is equipped with a water-cooled condenser at the top.
5. Repeat the extraction process twice and combine the supernatants into the same flask.
6. Boil the round-bottomed flask through an underneath heater.
7. Allow the distillation process for 3 h.
8. Read directly from the graduated oil collection arm the volume of the oil extracted.
9. Convert to the oil mass by multiplying the volume by the oil density.
10. The oil density can be measured with a hydrometer.

3.8 Surface Oil Content

1. Weigh 1.0 g of dry microcapsule and add it to 9 mL of petroleum ether or hexane in a glass flask.
2. Add to this 10 μg/mL of an internal standard.
3. Stir the mixture for 30 s at room temperature in a rotary shaker.

4. Centrifuge the mixture at 3000 rpm for 10 min.
5. Reserve the organic phase in a closed amber vial.
6. Analyze 2 μL of the organic phase in a Gas Chromatography (*see* **Note 16**).

3.9 Quantification Methodologies for Organic and Volatile Extracts

If the extraction method resulted in a volatile organic phase (with a boiling point inferior to the water) (e.g., ethanol, acetone, chloroform).

3.9.1 Gravimetry Quantification

1. Wash, dry, and weigh the flask to be used in the evaporator (*see* **Note 17**).
2. Insert the organic phase and couple the flask to the rotary evaporator. Use it under reduced pressure (*see* **Note 18**).
3. Alternatively, use a water bath to evaporate the solvent, but ensure the temperature does not exceed the degradation temperature of the analyte.
4. If the extracted material is still in the solvent, dry the extract under vacuum or by using a desiccator.
5. Weigh the evaporated extract and calculate the mass recovery by the difference with the empty flask.

3.9.2 Spectrophotometry

1. Dissolve the interested active compound (e.g., essential oil or vitamin) at a determined concentration (e.g., 30 μg/mL) in a solvent (e.g., ethanol, acetone, chloroform).
2. Scan the absorption spectrum of the active compound on a UV-visible spectrophotometer to detect the peak of the curve in the 190 to 500 nm range.
3. Make dilutions of the active compound in the same solvent (e.g., 10 to 200 μg/mL)
4. Read the absorption at the wavelength detected for the peak (*see* **Note 19**).
5. Plot the absorbance as a function of active compound concentration to have a calibration curve (Note: ensure the linearity of measurements with R2 => 0.98)
6. Add 3 mL of the extract to a cuvette to take an absorbance reading at the wavelength set for the active compound-solvent scan.
7. Take three absorbance readings of the supernatant of each system.
8. Use the equation from the calibration curve prepared to determine the concentration of the essential oil present in the microparticles produced (*see* **Note 20**).
9. Calculate the amount of active compound in the extract:

Amount (g) = [Concentration (mg/mL)]* Volume of extract [mL].

3.9.3 Gas Chromatography

1. Inject approximately 2 µL of the organic phase into a GC equipped with a polar packed column (e.g., 30 m × 0.53 mm; 1.5-µm film coating).
2. Adopt proper chromatographic conditions, e.g., flame ionization detector (FID) at 275°C with helium as the carrier gas; column head pressure 45 kPa; split injection; initial temperature 60°C with initial time of 2 min; program rate 8°C/min to 100°C, and 10°C/min to 200°C.
3. Adopt an external standard to calculate the quantity of a specific compound (e.g., limonene).
4. Prepare the standard substance in the same solvent as the sample in three concentrations (e.g., 25, 50, and 100 µg/mL).
5. Inject 2 µL of these standards into GC in the same conditions.
6. Use the average peak area to calculate the amount of the desired substance.

4 Notes

1. For dry particles, the dispersing media may be silicon oil.
2. To analyze the size of particles, prepare the glass slide with a few particles, trying to separate them from each other. Count about 150 particles to have a representative number of particles.
3. Insert the rule in the image.
4. If necessary, wash insoluble particles with distilled water three times to remove soluble substances from their surface before drying.
5. For delicate biological samples: Use Critical Point Drying (CPD) to replace ethanol with CO_2, avoiding surface tension damage.
6. Normally carbon adhesive tape or conductive silver paint.
7. If the pixel:length relationship is known from a previous measurement, you may directly type this information in the Set Scale window.
8. Typically, mesh sizes varied from 40 µm to 2.5 mm. For reference, AFNOR sieves are graded as follows: 40, 50, 63, 80, 100, 125, 160, 200, 250, 315, 400, 500, 630, 800 µm, and 1, 1.25, 1.6, 2, 2.5 mm.

9. The greater the number of sieves, the better the histogram will be achieved.
10. HCl is used to destroy the particles
11. Considering its optimal conditions of temperature, pH, and concentration.
12. Alternatively, a solution of alginate lyase and EDTA, or EDTA (0.1 M) and sodium citrate (0.2 M).
13. The amount of Solvent 2 may need to be adjusted based on the desired extraction efficiency.
14. Avoid emulsion formationemulsion formation due to agitation. The efficiency of liquid-liquid extraction can be influenced by factors such as the partition of the analyte in the solvents, the agitation time, and temperature. For highly viscous samples, pre-treatment with a sonication bath or ultrasonic probe may improve dispersion and enhance extraction. If phase separation is poor, consider using salt addition to improve the separation of phases or modify the solvent polarity.
15. Its important to know the partition coefficient of the core material in the solvent. Also, the volatility of the solvent should not be high.
16. Since the amount of surface oil is very small, an accurate quantification technique should be used.
17. The volume of the flask also determines the mass required for quantification. It is preferable to used smaller flasks due to the lower interference in the precision of the balance for quantify the final mass recovery.
18. The boiling point of the solvent determines the vacuum required.
19. The peak should be clear.
20. If the absorbance is greater than the curve, dilute it in the same solvent and re-read the absorbance.

Subject Index

A

Adhesion
- contact angle measurement 6

Agglomeration
- agglomeration index calculation 8–9
- fluidized bed 2, 6, 10
- hot-melt coating 1–3, 6, 9, 10

Alginate
- complex coacervation 81–82, 91
- ionic gelation 75–76, 144
- multilayer emulsion 94
- spray drying 30, 31, 33, 40

Analytical techniques
- Confocal Laser Scanning Microscopy (CLSM) 17, 151
- Differential Scanning Calorimetry (DSC) 42
- Dynamic Light Scattering (DLS) 30, 47
- Fourier Transform Infrared Spectroscopy (FTIR) 109–111
- Scanning Electron Microscopy (SEM) 150
- (X-ray Diffraction) 42, 47

Antisolvent processes
- SAS technique 114, 118
- nanoprecipitation 107–108, 111
- supercritical CO2 .... 114, 117, 119

Ascorbic acid (*see* hydrophilic active compound)
- solvent evaporation 52–53

B

Beeswax
- oleogels 122, 128

Bioactive compounds
- encapsulation efficiency 30, 32, 38, 82, 133, 155
- hydrophilic 38, 41, 43, 46, 71
- lipophilic 38, 43, 46, 61, 63, 71, 82, 85, 122, 124, 127–129
- stability 100–101, 103, 105

Biopolymers
- complex coacervation 81, 83–84, 91, 144
- polysaccharides 20, 81
- proteins 26, 81

C

Calcium chloride
- crosslinking agent 134

Carnauba wax
- hot-melt coating 2, 128
- oleogels 128

Chitosan 32, 78, 90, 95–96, 103

Coacervation
- complex coacervation 81, 83–84, 91, 144
- simple coacervation 81–82
- yield calculation 85

Coating efficiency
- hot-melt coating 8
- process yield 8

Confocal microscopy (CLSM)
- fluorescence labeling 94, 96, 151
- multilayer visualization 94, 96
- particle morphology 103

Contact angle
- adhesion 6
- goniometer analysis 2, 5
- wettability 100

Core material
- essential oils 19–20, 22, 26, 90
- hydrophilic materials 53–54, 138
- hydrophobic materials 26, 29, 52–53, 90
- vitamins 13, 19, 26

Crystallinity
- lipid nanoparticles 60, 62, 66, 71
- polymorphism 42, 47, 63, 69
- XRD analysis 42

Curcumin
- encapsulation 20, 46, 53, 76, 111, 144
- oleogel 129
- solvent evaporation 52–53

D

Density
- bulk 10, 22, 150
- characterization 22
- tapped 150
- true 150

Ana Silvia Prata and Louise Emy Kurozawa (eds.), *Biopolymers*, Methods and Protocols in Food Science, https://doi.org/10.1007/978-1-0716-5230-5,

Differential Scanning Calorimetry (DSC)
crystallization ... 37–38, 42, 46
melting point ... 42
thermal behavior ... 42
Droplet size
DLS ... 30
emulsion stability ... 29–30, 91, 96
static light scattering ... 30, 91
Drying methods
freeze-drying ... 127
moisture content ... 22
spray drying ... 25–26, 30–31
vacuum drying ... 139

E

Emulsification methods
high-pressure homogenization ... 64–65, 71, 96
rotor-stator ... 96
ultrasound ... 44
Emulsifiers
Lecithin ... 43, 70–71, 139–140
Span ... 70, 85, 129
Tween ... 70, 85, 112, 139
Emulsion stability
creaming index ... 30, 91, 102
Turbiscan analysis ... 91
zeta potential ... 91, 93
Emulsion-template oleogels
indirect approach ... 121–123, 126, 129
hydroxypropylmethylcellulose ... 126
sodium caseinate ... 129
Encapsulation efficiency (EE)
calculation ... 155
load ... 32
surface oil ... 30, 155
total oil ... 155
Essential oils
complex coacervation ... 81, 83–84, 91, 144
encapsulation ... 20, 46, 53, 76, 111, 144
extrusion ... 19, 22
spray drying ... 25–26, 30–31
Ethylcellulose
oleogels ... 122, 128
Extrusion
hot-melt extrusion ... 20
twin-screw extrusion ... 20–21

F

Fluidized bed
hot-melt coating ... 2, 128
minimum fluidization velocity ... 10
nozzle positioning ... 2, 7
Foam-template oleogels
freeze-drying ... 127
HPMC ... 126
oil absorption capacity ... 127

G

Gelatin
complex coacervation ... 81, 83–84, 91, 144
ionic gelation ... 75–76, 144
wall material ... 26, 146
Glycerol monostearate
oleogels ... 122, 128
Gum Arabic
complex coacervation ... 81, 83–84, 91, 144
spray drying ... 25–26, 30–31

H

High-pressure homogenization
emulsion formation ... 96, 101
lipid nanoparticles ... 60, 62, 66, 71
Hot-melt coating
fluidized bed processing ... 2, 6
lipid excipients ... 10
Hot-melt extrusion
essential oils ... 19–20, 22, 26, 90
non-expanded particle ... 20
Hydrogels
ionic gelation ... 75–76, 144
Hydrophilic active compound
interfacial polymerization ... 133, 137
ionic gelation ... 75
pickering emulsion ... 99
porous starch ... 13–14
solvent evaporation ... 51
spray chilling ... 37, 42, 46
spray drying ... 25–26
Hydrophobic active compound
complex caocervation ... 84
interfacial polycondensation ... 138
ionic gelation ... 144
multilayer O/W emulsion ... 78, 90
nanoprecipitation ... 107–109, 111
oleogel ... 121–129
pickering emulsion ... 99–101
porous starch ... 13
solvent evaporation ... 51
spray chilling ... 37–39
spray drying ... 25

I

In situ polymerization
melamine–formaldehyde ... 133–135
phase-change materials ... 134, 136

pH-responsive systems 134
Interfacial polycondensation
green encapsulation 138
polyurea 137
polyurethane 137
Ionic gelation
bead formation 79
dripping technique 75, 78, 79
external method 75–78
internal method 75–78
Isoelectric point
multilayer emulsions 90, 93–95
pH adjustment 32, 134
protein charge 95

L

Layer-by-layer (LbL)
electrostatic deposition 90
multilayer emulsions 90, 93–95
Lipid nanoparticles (LN)
high-pressure homogenization 64–65, 71, 96
Nanostructured Lipid Carriers (NLC) 60
Solid Lipid Nanoparticles (SLN) 60
solvent injection 61
Load
core material quantification 155
dry basis calculation 85
encapsulation efficiency 30, 32, 38, 82, 133, 155
Lutein
nanoprecipitation 107–108, 111

M

Maltodextrin
carbohydrate carrier 26, 32
wall material 26, 146
Melamine–formaldehyde
in situ polymerization 133–134
microcapsules
Microfluidization 66, 71, 96, 101
emulsion formation 96, 101
lipid nanoparticles 60, 62, 66, 71
Microscopy
CLSM 17, 103
ImageJ analysis 148, 153
morphology 103, 150
optical microscopy 55, 91, 103–104, 150
particle size analysis 55, 103
SEM 103
Moisture content
gravimetric analysis 156
vacuum oven method 86, 135, 149, 156
Multilayer emulsions
interfacial layers 90, 94
protein–polysaccharide systems 90, 93, 95
zeta potential 91, 93

N

Nanoprecipitation
antisolvent method 107
starch nanoparticles 110–111
Nile red 134, 151
fluorescence labeling 94, 96, 151
oil phase visualization 134, 151

O

Oleogelators
ethylcellulose 122
monoglycerides 122
sterols 122
waxes 122
Oleogels
direct approach 121–125
foam-template 122, 126–128
indirect approach 121–123, 126, 129

P

Particle Size Distribution (PSD)
polydispersity index 152
Pectin
ionic gelation 78
multilayer emulsions 90, 93–95
Phase inversion
lipid nanoparticles 60, 62, 66, 71
temperature control 68, 135
Pickering emulsions
contact angle 99–101, 103
particle stabilization 99–100, 102–103, 105
wettability 100
Poly(lactic-co-glycolic acid) (PLGA)
biodegradable polymer 52
solvent evaporation 52–53
Polymorphism
DSC 47
lipid crystallization 38
XRD 42
Polysaccharides
alginate 85, 90
chitosan 85, 90
gelatin 85
pectin 85, 90
starch 85, 100
Porous starch
enzymatic hydrolysis 14
probiotic encapsulation 13–16
Probiotics

encapsulation....20, 46, 53, 76, 111, 144
ionic gelation....75–76, 144
porous starch....14, 15
spray chilling....46
Proteins
casein....70
emulsifiers....33, 43, 63, 70–71, 123
pea protein....70, 85, 90
soy protein....70, 85, 90, 100
wall materials....26
whey protein....26, 70, 85, 90, 100

R

Release kinetics
controlled release....113
Rheology....29
emulsion stability....29–30, 91, 96
viscosity....29

S

Sieving
agglomeration index....8, 9
particle size analysis....55, 103
Solid lipid nanoparticles (SLN)
encapsulation....60
Solvent evaporation
double emulsions....52, 54
PLGA systems....52, 53, 55
O/W systems....51, 53, 55
W/O/W systems....52, 54
Spray chilling
crystallization....37–38, 42, 46
lipid microparticles....38
Spray drying
atomization....25–26, 30–31, 33, 40
encapsulation efficiency....30, 32, 38, 82, 133, 155
wall materials....25, 26, 28, 29, 32
Starch
nanoprecipitation....107–108, 111
porous starch....13–16
Surface oil
encapsulation efficiency....30, 32, 38, 82, 133, 155
gravimetric determination....155
solvent extraction....155
Surfactants
lecithin....140
Span....70, 85, 129
Tween....70, 85, 112, 139
Supercritical Antisolvent (SAS)
CO_2....113, 114, 119
co-precipitation....113–114, 117–119

T

Turbiscan
creaming....30, 91
emulsion stability....29–30, 91, 96

U

Ultrasound
emulsification....42, 101, 134

V

Viscosity
emulsion stability....29–30, 91, 96
rheological behavior....29
spray drying....26, 29, 32
Vitamins
encapsulation....20, 46, 53, 76, 111, 144

W

Wall materials
maltodextrin....26
polysaccharides....20, 81
proteins....26, 81
Whey protein
emulsifier....129
spray drying....25–26, 30–31

X

X-ray Diffraction (XRD)
crystallinity....42
polymorphism....42, 47, 63, 69
starch nanoparticles....110–111

Z

Zeta potential
electrophoretic mobility....86
multilayer formation....93
surface charge....86

The manufacturer's authorised representative in the EU is Springer Nature Customer Service Centre GmbH, Europaplatz 3, 69115 Heidelberg, Germany. If you have any concerns regarding our products, please contact ProductSafety@springernature.com

Printed and bound by CPI Group (UK) Ltd, Croydon, CR0 4YY
15/09/2026
02217005-0001